Environmental Democracy Facing Uncertainty

P.I.E. Peter Lang

Bruxelles · Bern · Berlin · Frankfurt am Main · New York · Oxford · Wien

Cécilia CLAEYS and Marie JACQUÉ (eds.)

Environmental Democracy Facing Uncertainty

Ecopolis
n° 16

Éditions scientifiques internationales
Brussels, 2012
1 avenue Maurice, 1050 Brussels, Belgium
info@peterlang.com; www.peterlang.com

ISSN 1377-7238
ISBN 978-90-5201-855-3

D/2012/5678/75

Printed in Germany

Library of Congress Cataloging-in-Publication Data
Environmental democracy facing uncertainty /
Cécilia Claeys and Marie Jacqué (eds.).
p. cm. -- (Ecopolis ; no. 16) Includes bibliographical references and index.
ISBN 978-90-5201-855-3 (alk. paper)
1. Environmental policy. 2. Environmental protection--International cooperation. 3. Democracy. I. Claeys, Cécilia. II. Jacqué, Marie.
HC79.E5E5765 2012 333.72--dc23 2012027871

CIP also available from the British Library, UK.

Bibliographic information published by "Die Deutsche Bibliothek"

"Die Deutsche Bibliothek" lists this publication in the "Deutsche Nationalbibliografie"; detailed bibliographic data is available on the Internet at <http://dnb.ddb.de>.

Table of Contents

Introduction

Cécilia CLAEYS and Marie JACQUÉ

Assistant Professors, Aix-Marseille University, France

The renewal of interest in environmental issues has spurred heightened interest in democracy as well. Declining biodiversity, climate change and sustainable development have progressively replaced nature conservation, fighting pollution and protecting the environment. This semantic shift may be interpreted as a simple change in nuance; but it appears that beyond the terms themselves, this evolution is the sign of a structural change in how issues are put on the public agenda and how related policies are shaped. Environmental issues are now defined, resolved and addressed on an international scale, within a scientific and administrative sphere comprised of the scientific community, national and international institutions and NGOs. The issue of democracy has become central in bridging the relationship between these different social bodies. Indeed, it must address new questions about the environment, marked by uncertainty and scepticism over the knowledge used in making choices and taking decisions. In this context, how should environmental decision-making be understood? On what rights is it based? On what conditions? Based on what knowledge and rationality should democratic debate take place?

This collective work takes an international look at the ties between environmental protection and democracy. Scientific and/or activist research has produced various terms to describe this phenomenon: ecological democracy (Morrison, 1999; Mitchell, 2006; Bourg and Whiteside; 2009); environmental governance (Arts and Leroy, 2006); and environmental democracy (Mason, 1999; Wilson 2006). Here we shall adopt the notion of environmental democracy in order to underscore the intrinsic connections between environmental and democratic reform. Indeed, decision-making processes related to the protection and management of the environment convey a will to reform the modern democratic model. These international, national and/or local reforms tend to share a common view: they encourage consultation and public debate around a central idea that sees (eco-)citizen participation as a new democratic ideal.

This book aims to provide answers to several questions. Is a post-modern era governed by the principles of participatory and deliberative democracy taking over from the modern anthropocentric era, which is organised, at best, around the principles of representative democracy? What would such a change mean, regardless of whether or not it is occurring? Is the environment a means to rethink democratic institutions or does the management of environmental problems reveal the weaknesses inherent to democratic systems?

At the international level, the 1992 Rio de Janeiro Earth Summit consummated the union between democracy and the environment. The connection between the environment and democracy indeed runs throughout the summit's declaration, which has since become a reference. Among other things, its article 10 stipulates that this connection needs to include the development of participatory democracy, which advocates the creation of mechanisms for consultation, to encourage citizen involvement and the participation of local populations in managing the environment:

> Environmental issues are best handled with participation of all concerned citizens, at the relevant level. At the national level, each individual shall have appropriate access to information concerning the environment that is held by public authorities, including information on hazardous materials and activities in their communities, and the opportunity to participate in decision-making processes. States shall facilitate and encourage public awareness and participation by making information widely available. Effective access to judicial and administrative proceedings, including redress and remedy, shall be provided (article 10, Rio Declaration, 1992).

Ever since then, national and international environmental management policies have taken a proactive approach towards implementing participatory democracy. At the European level, the Arhus Convention on "access to information, public participation in decision making and access to justice in environmental matters" (1998) confirmed the legacy of the Rio de Janeiro Conference and translated the connection between environmental management and the development of participatory democracy into Community policy. The integration of the participatory principle has involved, more or less, radical changes in different political traditions and national legal frameworks. Indeed, some countries have adopted the participatory principle more recently than others (Fourniau *et al.*, 2005). States in which democratic principles have struggled to see the light of day have often relied on (or given free range to) international organisations and NGOs, especially in Southern-hemisphere countries.

Consensual enthusiasm for such principles has made the connection between the environment and democracy seem obvious. And yet it is the fruit of a socio-historical process that we shall recall here in order to better understand the contemporary situation. To begin with, the connection between the environment and democracy – and even more so for participatory democracy – was not always obvious. Indeed, the first naturalist demands were voiced during the 19th century by a Western elite jealously defending the showcase of nature in a specific logic. An implicit part of protecting nature thus involved defending the interests of this educated elite. As such, local populations were at best ignored and at worst blamed for destroying nature (Kalaora and Savoye, 1986).

It was not until the late 1960s, with the shift from naturalism to environmentalism (Van Koppen and Markham, 2007), that the democratic argument was first employed. Driven by the new middle classes (see particularly Dunlap *et al.*, 2000; Mendras, 1988), concern for the environment distinguished itself from its naturalist past, notably by embracing protest. These new social movements, to borrow the term coined by Alain Touraine (1980), protested in block against the archaism of bourgeois morals and order, capitalist imperialism and the ravaging effects of modern science and technology. Staunchly opposed to the bourgeoisie's clientelist strategies, these new social movements, which filled the public sphere in the Habermasian and Goffmanian sense, criticised the closed-door negotiations of the traditional elite and revealed the limits of representative democracy (Ion, 2001).

Begun in the West, such concern for the environment gradually spread to all culture areas. In the Soviet Bloc, it played a role in challenging the communist dictatorships (Mandrillon, 2010). In the Southern hemisphere, it was involved in defending indigenous peoples from postcolonial imperialism (Guinier, 2004). Finally, the emergence of a middle class in the East and in the South contributed to the overall spread of concern for the environment or, to use Riley Dunlap's expression (Dunlap *et al.*, 2000), to spreading the New Ecological Paradigm (NEP), unless it was rather what Agarwar and Narain (1991) critically refer to as new environmental colonialism.

Faced with increasing environmental protest, governments often responded by integrating the criticism (Boltanski and Chiapello, 1999). Ministries of the Environment were created (e.g., in 1971 in France and 1970 in Canada) and have since legislated in favour of developing new cultural and environmental policies rethought in the light of participatory democracy. Embracing participatory democracy has allowed governments to address the demands made by new social movements to be heard and for transparency in the decision-making process, whilst luring these movements away from protest and towards making proposals. In this classic process of integrating criticism, the governments change whilst remaining themselves, whereas their opponents become partners (Lascoumes, 1994).

This virtuous compromise has, however, had several perverse effects. As such, the chosen or coerced opening of decision-making processes to an increasingly diverse group of social actors has created a paradoxical tension between institutional recognition of the diversity of vernacular and alternative knowledge, and a strategy to normalise such knowledge. Ministries of the Environment have been confronted with internal resistance to the polycephalous strength of the public from older and more powerful ministries in charge of the economy, industry, trade and/or labour (Lascoumes and Lebourhis, 1997). Moreover, the creation of consultation policies has raised the issue of cohabitation between representative and participatory democracies, which can be both competitors and complementary (Blatrix, 2009). Those involved are caught between the ballot as a binary means of expression and the opportunity – or was that difficulty? – to express their point of view during open debate in the public sphere (Chateauraynaud, 2007). In making reasoning the most powerful democratic tool, deliberation has placed knowledge and the ability to apply it, notably through linguistic habitus, at the centre of things. In doing so, consultation has sometimes taken the shape of democratic persuasion (Claeys-Mekdade, 2003). In terms of the deliberative process itself, co-decision making through the co-construction of knowledge long hoped for by some of those involved, as well as by some sociologists such as Michel Callon and Bruno Latour (Latour, 1999; Callon *et al.*, 2001), has not been as successful and reciprocal as expected. The hierarchy between scientific and vernacular knowledge has persisted, thus limiting the reciprocity of co-learning processes. Arguments about costs and economic input may be used to sever debate at any given moment, depending on the balance of power at play (Claeys, 2003; 2006).

The knowledge and co-construction processes at the root of participatory democracy have led to the emergence and recognition of groups of actors whose legitimacy is put to test in this context. The open and transparent approaches at its root have echoed and emboldened growing calls for citizen-based politics. Citizens' committees and juries are proof of the importance placed on this demand in the public sphere (Leach, Scoones and Wynne, 2005). Research into this process has examined the legitimacy of groups that, within the deliberative context, become spokespeople for public interest. Brandishing and mastering technical and scientific knowledge appear to be a vital condition for active participation in the democratisation of decision-making. Recognition of other types of knowledge, such as popular or vernacular knowledge about nature remains quite marginal; this has pushed the social sciences interested in environmental issues to promote the positive role such knowledge can play in managing resources (sometimes even becoming its spokesperson).

The opposition between expert and lay knowledge has been yet another impediment to the implementation of deliberative procedures and participatory democracy. Empowerment, which aims to strengthen the individual and collective ability to exercise power, is a concept now used in the environmental field (Bacqué, 2005). The knowledge and know-how of "local communities" have thus gained new clout in terms of the means available to claim and manage natural resources. The sociologists involved in their renewal and promotion have paid special attention to how power is organised, resources distributed and problems addressed. All of this interest has raised questions about how decision-making takes place and on what scale, in a context of increasingly regionalised public involvement and state withdrawal.

And yet, while social change is perceptible, the post-industrial (Touraine, 1971), post-materialist (Inglehart, 2008) and post-modern (Habermas, 1981; Giddens, 1990; Beck, 1992) society announced by observers looks more like a collection of jolts that are not quite strong enough to shake the foundations of the dominant, speculative and capitalist economic system in place. Confronted with such elusive social change, the same observers have introduced additional nuance and transitional models, such as the radicalisation of modernity showcased by Anthony Giddens (1990) or the two phases of (late?) modernity described by Ulrich Beck (1992). Or will compromise become a standalone model, as the tenants of ecological modernisation have suggested, underscoring the capacity of modernity's tools (e.g., science, technology, market economics) to work for the protection of the environment so long as the legal framework encourages them to do so (Spaargaren, 2000; Mol, 2000)?

These theoretical approaches converge around the importance placed on the reform of democratic institutions for addressing social and ecological problems. This has resulted in a shift in the social and political answers that can be brought to environmental problems. Rather than singling out and addressing the economic and social causes, contemporary public environmental policies have increasingly refocused on finding a means to adapt to and create social acceptance of the ecological limits of development.

This shift in decision-making about the environment has been accompanied by a ramp-up of uncertainty in how the relationship between nature and society is understood. Contemporary societies are often confronted with new risks, the result of scientific progress, which reduces the confidence placed in them to solve ecological issues. For Ulrich Beck and Anthony Giddens, this novel situation must foster new types of government and institutions that guarantee democracy. Bruno Latour's call for the implementation of a "Parliament of things" is similar in nature. Can democratic expression be improved by increasing the role of citizens in controlling technoscience? Which institutional forms provide a local space for expression on issues such as climate change, which are socio-politically defined and dealt with on a supranational scale? The different scales of mediation and the creation of deliberative procedures, as well as training and information about these issues, are all potential means for politically constructing a site for debate that is able to provide a social basis for environmental decision-making (Theys, 2003). Uncertainty about the limits of scientific and technological progress needs to be countered with a quest to "master" the future of societies; its cornerstone will be the renovation of political institutions. Given that it has been an international and cross-border problem from the outset, the environment is likely to be a privileged subject for analysing contemporary institutional changes. Analyses of experiments in local governance reveal an opposition between an optimistic vision and more pessimistic conclusions about the effects of democratisation on decision-making. Although the notion of governance is semantically imprecise and polysemic, it represents a new institutional framework in which debate over environmental issues can be tested. Without a doubt, one benefit of having multiplied the sites of participatory democracy is the challenge it poses to the monopoly of scientific expertise in decision-making.

The creation of deliberative mechanisms and means for citizen involvement in managing the environment thus appear to be social and political innovations that have in part addressed the difficulties faced by environmental policy. A classic legal, statutory and sector-based arsenal has been aimed at issues characterised by their versatility and complexity. A more critical (or at least sceptical) position has emerged which questions the "environmental efficiency" (Salles, 2006) of governance measures. Indeed, the contemporary scope of such "democratic concern" seems, through a counter-productive side effect, to have almost obscured the importance of protecting resources. If democratic effort has given the protagonists a stronger voice, what is the result following debate over how to improve protection of the environment? And how can such environmental improvements be objectively assessed in an international context of scientific uncertainty? The theoretical and empirical limits of major global concepts (biodiversity, climate change, sustainable development, etc.) have encouraged an increased blurring of the lines (Latour [1999] would say hybridisation) between knowledge and ideology and science and politics. Faced with such uncertainty (notably scientific), the response that consists of resorting to the principle of precaution, particularly popular in the 1990s, has quickly shown its limits (Larrère C. and Larrère R., 2000; Bronner G. and Géhin E., 2010) since it results in fluctuation between no decision and over-protection, once again raising inevitable questions about the decision-making process.

In this respect, can the participatory logic not also lead to a weakening of critique within the democratic process which, to express itself, needs to accept the institutional framework of deliberation? Behind this paradox lies a fundamental challenge for the relationship between democracy and the environment: that of "making hard decisions" in a "weak" political context (Roqueplo, 1997) in which individual freedom – often associated with or reduced to producing and consuming – is the most important value. The rise of environmental democracy has tended to shift debate about the causes of ecological problems towards the construction and legitimation of frameworks in which debate should take place.

By confronting international research with local and national experience, the authors in this book – of nine different nationalities – have different views about environmental democracy. Each chapter focuses on one component in the complex process of connecting the environment and democracy, providing information about the actors, bodies and issues, but also more widely about the paradoxes and dilemmas of this world in the making, caught between reproducing the old and creating something new, between the individual and society.

Hilary Tovey underscores how the individualisation process in modern society has challenged the exercise of citizenship constructed within the republican framework of the nation state. Through a sociological analysis of different types of environmental activism in Ireland, ranging from organised collective action to individual commitment in the private sphere, the author examines types of environmental protest that co-exist more than they oppose each other. Through a comparative approach to the sociology of collective action, the environment and eating habits, Tovey argues that the development of "eco-citizen" behaviour is part of a process of lifestyle moralisation that has a limited contesting effect. This contribution pushes us to reflect on the importance of different types of public expression within the scope of modern democracies.

Increasingly faced with problems, different types of citizens and eco-citizens sometimes need to bond together to create a public, particularly when they participate in deliberative processes. Luigi Pellizoni shows how in this context environmental democracy has redefined the notion of public. Starting with an overview of international sociological literature, the Italian author describes the possible paradoxical effects implicit in this new definition. He underscores the tension and interdependence between individual interests and the common good and as a corollary between the private and public management of environmental assets. The author looks at conceptual contributions from economics, political science and sociology, and points out the role environmental issues have played in redefining the political foundation of Western society. Without disregarding the possible illusory effects of "greenwashing", Pellizoni convincingly argues for democratic renewal in favour of what he calls the third generation of environmental governance, which will allow a broader public sphere to emerge as long as two favourable conditions are present: reciprocity and openness.

Raymond Murphy's contribution looks at citizens torn between individual and collective action, people who form a heterogeneous and sometimes paradoxical group that expresses itself in a political and scientific context undergoing major change. We have increasingly begun questioning our knowledge of Nature/Society interactions. Here, the author takes a closer look at how decisions are taken and public policies conducted in a new paradigm in which scientific progress leads only to partial knowledge. We are invited to embrace the counter-intuitive idea that increased knowledge about nature will not necessarily reduce our ignorance. One consequence of this is that public decision-making has become increasingly complex and has fundamentally changed the way knowledge is handled in decision-making. Murphy uses the example of dealing with climate change and the exceptionally bad weather that has recently affected Canada and the United States to show how decision-makers have built uncertainty into their choices. By upsetting the boundaries between expert knowledge and social experience, disasters play a central role in reshaping democratic deliberation on the one hand, and in improving warning mechanisms on the other.

Increased reliance on scientific knowledge is a major characteristic of environmental democracy. Such knowledge, however, is both ample and incomplete; alone, it is insufficient for resolving debate. That is at least one of the reasons for opening the debate to other so-called popular or vernacular types of knowledge. In France and Germany, two teams have developed applied research projects to encourage the taking of such vernacular knowledge into account in decision-making processes. Martina Schäfer and Tina Boeckmann bring both a theoretical and methodological element of response based on an analysis of "social-ecological research", an innovative approach to research-action. While the interdisciplinary approach used in the research projects studied is a necessary condition for integrating popular knowledge, it is not necessarily sufficient. Indeed, the authors underscore that the type of knowledge expected from research and the objectives set by the different stakeholders involved are also determining factors. Participation and its implementation are thus a more or less central issue depending on the intended purpose of the research. In France, Audrey Richard and Olivier Barreteau use an experimental game to promote dialogue between the different types of knowledge present in a single territory. Using the hybrid forum framework (Callon *et al.*, 2001) and "regimes of engagement" (Thévenot, 2000; Boltanski and Thévenot, 1991), the authors emphasise the plurality of knowledge involved, its (re-)formulation and (co-)construction. They underscore the connection between three major regimes of engagement: moral capacity, strategic capacity, and feelings and emotions. The case study presented focuses on collective management of the Lentilla and Llech river basin. With its dry Mediterranean climate, potentially conflicting and competing water uses are common in this area and create tension between the different needs for drinking water, agricultural water and water for fishing and tourism. The methodological originality of this research stems most notably from its use of an experimental game whose goal is to help the different parties involved reach a compromise on their own. Here the sociologists are mediators who supply sociology's analytical tools to foster debate.

In other more conflicting areas, however, sociologists are more critical. Importation of the Western model of environmental democracy has indeed raised new problems. Between economic pressure caused by the desire for local resources and the trouble NGOs have in shedding their ethnocentric vision of environmental issues, the local populations in Southern Hemisphere and Eastern countries are faced with a unique set of issues. Maria Tysiachniouk and Errol Meidinger show how the creation of deliberative processes around forest management in Russia has contributed to a Westernisation of not only institutional thinking but also of the collective imagination of nature. Using case studies from the Pskov and Preluzie forests, the authors analyse the alliances and opposition that exist between the local economy still marked by the Soviet era, international firms in search of new markets, international NGOs active and experienced in networking, as well as local populations with no experience in participatory democracy. In this context, consultation at best resembles an information campaign and, at worst, a lobbying strategy for the sustainable exploitation of a forest in the concrete and questionable form of a partnership between foreign companies and international environmental NGOs.

In Brazil, the struggle of small farmers to maintain their lifestyle has resulted in the innovative experiences and resistance strategies analysed by Angela Duarte Damasceno Ferreira and her team. This contribution places democracy at the centre of its analysis, not in the dominant terms of participation but through the creation of community-based approaches to reclaiming power and decision-making at the local level. The authors underscore that it is at the local level that we can best analyse and understand the resistance of small farmers to the hegemonic powers – whether economic or environmental – that marginalise them. The concept of empowerment is examined and used to explain this process. While it is a technique that can enrol small farmers in new types of participation, it is also a framework of thought to describe the different types of reclaiming by small Brazilian farmers, not only of power but also of their identity. Thus, what appears to be at the heart of this "local democracy" in action is the recognition of knowledge and habits often described as "traditional" and which, through the examples given, are shown to actually be a "modern alternative" that allows social, economic and ecological development to be interpreted in sustainable terms.

Finally, to open up much more than to conclude this vast field of research, Pieter Leroy provides a reflexive synthesis of contemporary research in the social sciences related to environmental democracy. To do so, he uses a structured typology based on two main inputs (mode of analysis and level of analysis), which allow for a critical reading of the framework behind the sociology of the environment, its claims and recent changes. Underscoring the enthusiasm first garnered from the optimism of analyses influenced by the sociology of translation, the author has observed a recent return to more critical sociology, which also happens to be more pessimistic. Drawing lessons from its analyses of environmental democracy, the sociology of the environment thus appears to now be seeking to find a heuristic balance between almost naive optimism and hopeless deconstruction.

References

Anil A., Sunita N., *Global Warming in an Unequal World: a Case of Environmental Colonialism*, New Delhi, Centre for Science and Environment, 1991.

Arts B., Leroy P., *Institutional Dynamics in Environmental Governance*, The Netherlands, Springer, 2006.

Beck U., *Risk Society: towards a New Modernity*, London, Sage, 1992.

Blatrix C., "La démocratie participative en représentation", *Sociétés contemporaines*, 74, 2, 2009, pp. 97-119.

Boltanski L., Chiapello E., *Le nouvel esprit du capitalisme*, Paris, Gallimard, 1999.

Callon M., Lascoumes P., Barthe Y., *Agir dans un monde incertain. Essai sur la démocratie technique*, Paris, Seuil, 2001.

Bronner G., Géhin E., *L'inquiétant principe de précaution*, Paris, PUF, 2010.

Chateauraynaud F., "La contrainte argumentative. Les formes de l'argumentation entre cadres délibératifs et puissances d'expression politiques", *Revue européenne des sciences sociales*, XLV, 136, 2007, pp. 129-148.

Claeys-Mekdade C., *Le lien politique à l'épreuve de l'environnement. Expériences camarguaises*, Bruxelles, PIE Peter Lang, 2003.

Claeys-Mekdade C., "La participation environnementale à la française : le citoyen, l'État et le sociologue", *Vertigo, revue électronique*, 3, 7, 2006, URL: http://vertigo.revues.org/8446; DOI: 10.4000/vertigo.8446.

Defrance J., "Donner la parole. La construction d'une relation d'échange", *Actes de la recherche en sciences sociales*, 13, June 1988, pp. 52-66.

Dunlap R. *et al.*, "Measuring Endorsement of the New Ecological Paradigm: a Revised NEP Scale", *Journal of Social Issues*, 3, 50, 2000, pp. 425-442.

Giddens A., *The Consequences of Modernity*, Cambridge, Polity, 1990.

Guignier A., *Le rôle des peuples autochtones et des communautés locales dans le développement durable: figurants ou acteurs ?*, Limoges, Presses universitaires de Limoges, 2004.

Habermas J., "New Social Movements", *Telos*, 49, 33-7, reprinted in Redcliff M., Woodgate G., *The Sociology of the Environment*, V, III, Aldershot and Brookfield, UK, Edward Elgar, 1995.

Inglehart R., "Changing Values among Western Publics, from 1970 to 2006", *West European Politics*, 31, 1-2, January-March 2008, pp. 130-146.

Ion J. (ed.), *L'engagement au pluriel*, Saint-Étienne, Publications de l'université de Saint-Étienne, 2001.

Kalaora B., Savoye A., *La forêt pacifiée. Sylviculture et sociologie au XIXe siècle*, Paris, Fayard, 1986.

Lascoumes P., *L'éco-pouvoir. Environnements et politiques, Les associations au risque de la démocratie participative*, Paris, La Découverte, 1994.

Lascoumes P., Lebourhis J.P., *L'environnement ou l'administration des possibles. La création des directions régionales de l'Environnement*, Paris, L'Harmattan, 1997.

Latour B., *Les politiques de la nature, ou comment faire entrer les sciences en démocratie*, Paris, La Découverte, 1999.

Larrère C., Larrère R., "Les OGM entre hostilité de principe et principe de précaution", *Cités*, 4, 2000, pp. 34-46.

Leach M., Scoones I., Wynne B. (eds.), *Science and Citizen: Globalization and the Challenge of Engagement*, London, Zed Book, 2005.

Mandrillon M.H., "La Russie face à l'environnement : une ambiguïté durable", in Favarel-Garrigues G., Rousselet K. (eds.), *La Russie*, Paris, Fayard, 2010.

Mendras H., *La seconde Révolution française, 1965-1984*, Paris, Gallimard, 1988.

Mol A., "Globalization and Environment: between Apocalypse-blindness and Ecological Modernization", in Spaargaren G., Mol A., Buttel F. (eds), *Environment and Global Modernity*, London, Sage, 2000, pp. 121-150.

Roqueplo Ph., *Entre savoir et décision : l'expertise scientifique*, Paris, Paris, INRA, 1997.

Salles D., *Les défis de l'environnement : démocratie et efficacité*, Paris, Syllepse, 2006.

Spaargaren G., "Ecological Modernization Theory and the Changing Discourse on Environment and Modernity", in Spaargaren G., Mol A., Buttel F. (eds.), *Environment and Global Modernity*, London, Sage, 2000, pp. 41-72.

Touraine A., *The Post-Industrial Society. Tomorrow's Social History: Classes, Conflicts and Culture in the Programmed Society*, New York, Random House, 1971.

Touraine A. *et al*, *La prophétie anti-nucléaire*, Paris, Seuil, 1980.

"Collective" and "Personal" Environmentalism

Implications for Democracy of the Greening of Citizenship

Hilary TOVEY

Fellow Emeritus, Trinity College Dublin, Ireland

This chapter discusses different ways in which people practise citizenship through their environmental activism. I try to link that issue to some recent debates in Social Movements theory, in environmental literature and in the sociology of food, which address the changing forms taken by political activism in an increasingly individualised society.

Discussions of citizenship generally start by noting two points: first, that the rights attributed to citizenship have expanded over time, from economic, to political, to social (and perhaps now also environmental) rights; and second, that different political philosophies understand the meaning and practice of citizenship in different ways. Marshall's argument about the historical elaboration of individual rights appears to fit best within a "liberal" model of citizenship. Leach, Scoones and Wynne (eds., 2005) identify liberal perspectives on citizenship as emphasising the granting of universal rights to citizens on the basis of their rational capacity to act on their own interests; in this model, the practice of citizenship involves the exercise of rights by citizens, particularly rights to participate through electoral democracy. The state also has a protective role towards its citizens, reducing uncertainties that arise out of economic processes and, increasingly, attempting to manage risks to safety from the industrialisation of food and from environmental exploitation, for example. As liberal theory understands citizens as rational individuals, it has difficulties in addressing cultural groups and also in recognising the possibilities of plural rationalities and knowledge within a given society. "Communitarian" perspectives emphasise "the notion of the socially embedded citizen and membership of a community" (Leach *et al.*, 2005: 23); the practices of citizenship are those that prioritise the common good over individual interests. Finally, "civic republican" perspectives address individual citizens as rational actors pursuing interests, but locate them within collectivities; recognising the division of interests and of power in society, it is expected that rational actors will form groups and pursue their claims collectively in the political process. The practice of republican citizenship is based on recognition by individuals of "obligations to participate in communal affairs" (2005: 24), using either representative democracy structures or alternative means. Civic republicanism offers an "agonistic" vision of citizenship in which struggles against domination and the abuse of power will never cease to be necessary.

Leach, Scoones and Wynne discuss these different perspectives on citizenship because they are interested in the interrelationships between notions of citizenship and human agency, forms of knowledge, and scientific and policy discourses on the environment. They suggest that scientific or "expert" discourses carry within them implicit understandings of what citizenship means and how it is appropriately practised. Expert environmental or risk discourses assume a liberal perspective on citizenship, attempts to manage environmental problems in a more "participatory" way often proceed implicitly from a civic republican understanding of citizenship. Their focus, then, is on how "citizens" are constructed by policy processes, and the constraints that follow on from what is regarded as appropriate citizen involvement in managing the environment. While this produces many very illuminating analyses in the book, it does not have much to say about how citizens themselves may respond to the conceptions of citizenship embedded in policy discourses and expert knowledge. Who accepts and who resists the "liberal" version of citizenship promulgated in expert discourses about environmental issues, and under what circumstances? And how is that related to the different ways of practising environmental citizenship?

In this paper, drawing on material collected in the Irish context, I set up a contrast between "individualised" – or, as I call them here, "personal" forms of environmental activism – and "collective" forms and suggest that these are two distinctive ways in which environmental citizenship can be practised, which produce and reproduce different understandings of citizenship and have different consequences for democratic engagement. I want to preface that discussion, however, with diversions into two different bodies of literature, first looking at some recent changes in the theorisation of social movements, and second, drawing on recent debates within the sociology of food over the form and nature of contemporary "food consumer politics".

Re-thinking collective action

Liberal perspectives on citizenship have been "co-produced" (Leach *et al.*, 2005: 22) with the historical rise of liberal capitalist societies.

Many important social theorists today argue that the social world that contemporary liberal capitalist democracies construct for their subjects is an increasingly individualised one (for example, Giddens, 1991; Beck and Beck-Gernsheim, 2002; Bauman, 1999). The thesis of individualisation, along with post-structuralist challenges to essentialist conceptions of collective identity, provides a significant challenge to theorists of collective action, particularly Social Movements theorists. Bauman (1999), for example, describes an individualised world as an "anti-collective" world: it is a world in which developing and maintaining shared lines of group action becomes an increasingly difficult and rare accomplishment. Habermas' concept of "civic privatism" suggests that the subjects of this world tend to retreat into their own private networks and relationships, or to become more concerned with personal and domestic projects than with public and political issues. We could read such accounts as suggesting that the process of individualisation brings about not only the end of shared, collective action, but the end of citizenship itself as a practice of active engagement with the wider society in which one lives.

Among social movement theorists, the individualisation challenge has induced some interesting attempts to re-think the idea of "collective action". Shared lines of group action, they suggest, do persist in contemporary society, but they are profoundly altered in form. For example, Paul Lichterman (1996), in an early discussion on the issue, compared working class or black environmental justice movements in the USA with middle class environmental activists. He found that the Environmental Justice groups he studied had a culture of committees, elections and community representation through formal organisation, whereas the middle class activist groups used a discourse of personal commitment and individualised decision-making. Lichterman characterises them as collectivised through a "politics of selves", rather than through a collectively shared and communal identity, and suggests that they represent the form that environmental mobilisations will increasingly take in the future. McDonald (2004) draws on Boltanski and Thévenot (1991) to argue that social movement theory has historically operated with a "civic-industrial" convention or model of collective action, which emerged within a particular historical context. The formation of European nation-states and the development of democracy shaped a "civic grammar" in European culture, which prioritised the moral worth of the collective over that of the individual. It regarded the private sphere of interpersonal relationships and commitments with suspicion, because they could impede the citizen from acting to realise the "general will" or "general good". Within social movements theory, McDonald argues, this "civic-industrial grammar" has led to an assumption that successful collective organisations must create for their members a collective identity in which the "I" is submerged in the "we", so that the actor can be transformed into "a serviceable agent" of the movement. But this model of a social movement no longer coheres with the actors' own subjective experiences of participating in collective action; the separation of public and private spheres, so central to the civic-industrial paradigm, becomes meaningless in a situation where participants construct shared lines of group action out of their close personal knowledge of each other, interpersonal trust and their recognition of personal differences among actors rather than attempting to construct sameness. Collective action persists, but its "civic grammar" is no longer that of a structured and hierarchical organisation, more that of an "affinity group" – a group of "like-minded" friends and acquaintances who work together to plan and carry out lines of action that are complementary to the goals of the wider "movement" but are initiated at group and individual level. (See also Farro [2004] on how people in the anti-globalisation movement define themselves as movement "members".)

Such attempts to revise how "collective action" and "social movement" are conceptualised in social movements theory suggest that we should not try to draw sharp boundaries between "collective" and "individualised" practices of citizenship. They imply that it may almost be a matter of contingency (happening to identify an accessible group of "like-minded persons", for example) whether one practises citizen protest or resistance as an individual, or in a group. Dubet and Lustiger Thaler (2004) summarise the shift as leading students of contemporary collective action towards a "subject-centred" theory of social movements. This would focus on understanding how subjects bring to the task of constructing "shared lines of action" their own sense of personal distinctiveness and individual goals.

Debates around "personal environmentalism"

We can locate similar shifts towards a more individualised approach to activism in some recent literature on environmental activism. Historically, environmental sociology has assumed that we can make a distinction between "environmentalism" and "societal greening". "Environmentalism" is used to refer to collective, social movement-based or organised activism on behalf of nature. It is understood as articulating some elements of social critique or opposition to the state and its associated experts and authorities, and represented as having been at its strongest during the 1970s and 1980s, then subsequently declining. "Greening", on the other hand, refers to the adoption, by individual citizens and by institutions, of "good environmental practices" in their domestic or commercial lives. Its current expansion indicates that recognition of environmental impacts is increasingly being "integrated" or "internalised" into everyday productive and reproductive practices.

Eder (1996) suggests that greening has become the "mainstreamed" form of ecological action over the past two decades. It can be seen as an outcome of the diffusion of Ecological Modernisation ideas and policy practices, which argue that the ecological modernisation of states stimulates "self-regulation" within civil society (Spaargaren *et al.* [eds.], 2000), and that "environmental governance", cooperation between the state and civil society, and citizen and corporate education, are more effective ways of moving towards sustainability than the oppositional and critical tactics of earlier collective environmental movements. The goal of environmental action moves from policing the state to informing – and (see Leach *et al.* above) forming – the citizen.

On the other hand, Andrew Jamison (2001) theorises "greening" in a rather different way: he puts it under the heading of "personal environmentalism", and includes this in a typology of environmental activisms where it represents one variant alongside three others: the "community", the "militant" and "professional" types. Jamison argues that the community and professional varieties of environmentalism engage in political activism, but the militant and personal types are better seen as forms of moral practices – the first in public, the second through private lifestyle and consumption practices. From his perspective, then, personal environmentalism is not an individualised replacement for earlier collective forms, but rather just one form that activism may take within the larger "environmental movement" found in contemporary societies.

The individualisation thesis, as I suggested earlier, poses a problem for social theory: how to categorise and position the moral practices of individual citizens in relation to social movement activism. The problem has surfaced not only in social movements theory and in environmental sociology but also recently in the sociology of food (which is of course not unrelated to environmental topics). So, I briefly digress into that arena to sketch out how it has been addressed there. The issue that food sociologists have been debating (see e.g. Buttel, 2000; DuPuis, 2000 and 2002; Goodman and DuPuis, 2002) is about the role that consumers can play, if any, in resisting or changing the contemporary global food system. Within the sociology of agriculture and agro-food, sociologists have tended to focus on food producers and retailers as active agents, but to assume that consumers are largely passive, manipulated and/or ineffectual in achieving change. But, increasingly, this is recognised as inadequate as a way of exploring the development of "food systems" or "food chains". Consumption has to be brought back into the picture and recognised as a distinctive location of agency. In turn, this has generated a discussion of the notion of "consumer politics": can this term only be applied when consumers are acting as a collectivity, or can politics be practised by individual consumers making their food choices in their local supermarket or food store? Buttel (2000) identifies consumer politics with collective consumer action within a broader social movement, and argues that consumer politics, in this sense, has shown itself to be weak and ineffective as a form of resistance to the power of big business and transnational institutions such as the WTO. Those who understand consumer politics in a more individualised way, however, assess it as a growing and dynamic force within food market relationships.

Goodman and DuPuis, among others, refer to the political action of individual consumers as a "new" politics of food. It is "new" in two senses: first, they are positing that there has been a decline in or a movement away from the type of consumer politics that involves relations between the state and organised consumer interest groups (again, seemingly, a movement away from the "civic-industrial" model of political action by citizens); and second, because, they argue, there is a new interest among consumers in issues other than price when making choices about food purchase – consumers are increasingly concerned about food quality and safety, fair trade, social justice, animal welfare, environmental impacts and the use of genetically modified organisms in food production. Such concerns are understood as part of a "contested bio-politics of agro-food networks" (Goodman, 1999), which transcend and link the spheres of production and consumption. For example, in her book on milk within American history and culture, DuPuis (2002: 215) rejects the view that organic milk consumers "engage in a neo-liberal form of 'false' politics, based on an agglomeration of individual rational decision-making and catering primarily to the status-conscious upper class". In her view, this "new form of consumption" is "explicitly political", although "less organised": "A consumer who is not a member of a Social Movement can still act politically if she or he takes into account the various political claims about a product in the process of making a purchase" (2002: 228) – claims that are made both in the public sphere and within the consumer's own social networks or "community of practices". This "process of taking in claims but not necessarily espousing any of them" is one she describes as "reflexive consumption". Thus the "new politics of food" is new in a further sense as well, in that it is practised by a new type of consumer: a reflexive and active actor who "is not a political activist" but who "thinks about consumption" (2002: 233), and in so doing can exercise "a powerful force" (*ibidem*: 232) for change within the food industry. A similar argument is made in the introduction to Wright and Middendorf ([eds.], 2008).

There are a number of issues one could raise about such analyses, but here I just want to highlight one. It concerns the complacency with which many contemporary sociologists appear to view the (supposed) transition from collectivised to individualised citizenship practices. Both the recent attempts to re-think collective action, and recent theorising of the "citizen consumer" as a political actor, are converging to a point where collective citizen action is seen as marginal or unnecessary. DuPuis, for example, appears to assume that the choices of the "reflexive consumer" are an adequate replacement for declines in collective activism, although the evidence she gives to support her view of them as a "powerful force" for change, whether in food, animal welfare or environmental management practices, is weak, to say the least. I argue here instead that a form of politics that takes place through private choices within the market is compromised as a terrain on which to learn or demonstrate practices of citizenship.

"Collective" and "personal" environmentalists in Ireland

During the spring and summer of 2004 we carried out a series of interviews with Irish environmental activists,[1] primarily to understand how they came to be formed as activists and their route into membership of environmental activist groups. We located 33 people to interview by contacting the group in which they were participating, and we called these "collective" environmentalists. They came from 23 different groups or organisations, at a national and local level. Some were more formalised, while others were quite informal. They were pursuing a variety of different environmental concerns – from wildlife and natural resource protection to attempts to close down the Sellafield nuclear power station in the UK – and from Green Party to anti-globalisation politics. We then interviewed five people who were known to be deeply engaged in environmental practices in their domestic lives but were not participants in any collective group. They were included because I wanted to explore further the notion of a "green citizen" or "personal environmentalist". It is important to emphasise that the sample is very small. All those interviewed were women living in middle class areas of Dublin and I am taking a risk with generalisations about the findings from this group.

As the discussions by Jamison, Lichterman and McDonald predict, in some respects the boundaries between these two groups of "collective" and "personal environmentalists" could be treated as porous or unimportant. For one thing, they share a great many lifestyle practices in-

[1] The research was funded by the Environmental Protection Agency (Ireland), under the National Development Plan 2000-2006, as part of a three-year project investigating "Environmental beliefs, values and behaviour in Ireland". I want to thank my two research assistants, Noelle Cotter and Adele McKenna, for conducting the interviews, under often quite difficult conditions. A full report on the research can be found in Tovey (2007).

tended to lessen their environmental "footprint" on the world. All the personals, and nearly all of the collectives, do the following: separate and recycle their domestic waste; try to consume in an environmentally friendly manner (for example, using household detergents or paints that contain no dangerous chemicals); seek out food that is not pre-packaged and sometimes organic; garden organically; and try, without much success, to cut down on the car as their mode of transport. In both groups we find individuals who bring their environmental concerns into their working lives. One of the personals is an interior design consultant; she does not advertise herself specifically as an eco-consultant, but uses her work as an opportunity to nudge clients into using ecologically-friendly materials and practices in their houses. Another manufactures and distributes recycled paper products. Among the collectives, we find, for example, organic farmers and homesteaders, a lecturer in environmental engineering (who abandoned a career in the corporate world because he did could not accept the environmentally destructive engineering projects he was being asked to carry out), and the owner of a business that designs and installs ecological systems for human waste disposal.

A second relevant point is that the dominant form of organisation among the collective activists we interviewed (and in this, we would argue, they were quite typical of the broader environmental movement in Ireland) does not follow the "civic-industrial model" referred to above; it tends to be quite informalised and to exhibit a lot of the characteristics of an "affinity group". People bring to their groups a clear sense of their own individuality and personal responsibility as an environmental actor, and they understand membership in the group as a matter of what they can individually contribute to it, rather than as a matter of accepting and following group identity, norms and rules. This might suggest that the fact that none of the personals was affiliated with any environmental networks or associations is just a contingency: living in a large, anonymous urban conglomeration, they just happened not to be in a position where they would be drawn into such a group, whereas if they were living in a rural town or village, for example, they probably would not have been able to escape induction.

However, the personals' own accounts of their position challenge this suggestion. They offer very clear reasons why they do not belong to an environmental group. These fall into two categories. First, most have accepted standard negative stereotypes of the environmental activist as "the brown rice and sandals brigade", the ageing hippy with a grey ponytail, people who wear parkas, and so on, and express distaste at the thought of being themselves labelled in such ways. As one put it: "Well, there is environmentally aware, and then there is environmental lunatics." Second, they dislike what one called "mob mentality". For example:

> Individuals are different, individuals can be educated, can educate themselves, can formulate their own opinions…What makes me nervous is when people are in collective sort of, you know, this mob thing can get very much influenced by different things, be it racism, be it you know the battery chicken, be it whatever.

Or as another put it: "I would have quite eclectic views on a lot of things, too, so I don't necessarily want to wear the badge of one particular, you know, group of people."

For these personal environmentalists, joining a group submerges one's personal identity in a collectivity that cannot be relied on to act in a rational, reflective way. Only the individual, in their view, can act on the basis of the acquisition of good, scientific or authoritative knowledge and can reflexively incorporate such knowledge into a moral line of action. When individuals allow themselves to become part of a group, they inevitably yield up their personal self and personal morality to "mob" rule.

The environmental practices of the personals, particularly perhaps their concern with managing their domestic waste, exemplify what McDonald (2004) calls "a moral cultivation of the self", or Giddens (1991) "a project of the self". In sorting, separating and recycling waste, they are cultivating an experience of personal difference, or of intensified personhood. Throughout the interviews with them, phrases such as "my hands are clean", or "I am doing my bit" recur:

> I would think at the end of the day it [environmental involvement] is about making choices, and I would think that if you're going to be aware of the fact that you have a choice you have to try and inform yourself… It sits easier on my conscience that I am doing the best for my children and my micro-climate and environment.

The emphasis on personal responsibility, feelings of guilt and the importance of hygiene, order and method suggest a process of cultivation of the identities of housewife, mother and domestic worker, which elevates their status into a "vocation" or moral calling (Weber):

> I am an irritating customer for shops, because I do go up and speak to the manager and I do say why do you not have Fair Trade, or why do you not have organic this, and this isn't good enough… I do think the consumer is king in today's society and if a demand is seen to be there they will actually do it.

In some of the interviews with the personal activists, the discourse of personal morality overlaps with a discourse about the self as a consumer. Consumers are represented as people who exercise personal choice, and they have an obligation to inform themselves so that their choices are moral and responsible. "I would be part of a wider group of people who would be trying to help the environment, and the bit that I'm doing is helping a wider group of people who are doing the same thing and feeling the same way." The way to bring about social and environmental change, for these personals, is not through collective action but through aggregated individual expression of personal concerns and demands.

Interviews with the collective environmentalists, however, suggest that there are significant differences between aggregated individual action and group action that develops through interaction and a shared history of experience among the participants. While the "subject-centred approach" to social movements is very illuminating for exploring how collective actors act, over-reliance on it can lead us to forget that groups are not only formed by acting subjects, groups can also form those subjects in significant ways.

I only have space here to sketch out some of the relevant features of the collective activists. First, these actors are characterised by a strong concern about society as well as about the self. They do also morally cultivate their "selves", and they do negotiate their identity as an individual person in contexts where others seek to label them as members of a particular group. But this is only part of their story. Often starting out as a group of like-minded neighbours and friends concerned about some aspect of their environment (wanting to protect a local ecological site or a wild species, for example), and who expect, perhaps naively, that their concerns and expertise entitle them to participate in environmental governance processes, they go through a dramatic learning process about the society they live in and how it operates. They learn that to realise their goals requires not just individual change but social change. We might say that they learn to develop a concept of society as a set of structures and relational processes that are primarily organised around the holding and use of power. Organising protests, meetings and lectures, bringing information to the attention of politicians and state agencies, operate as a sort of rite of passage that transforms their collective understanding of the world they live in. In particular, it generates a collective distrust of the political system and huge anger at the cynicism, disinterest and dishonesty it reveals to them. A few quotes help to illustrate this:

> We had one waste of a meeting with a Minister of State at the Department of the Marine which was so bad, two of the four of us got up and left, we were just being railroaded and abused, told we were stupid and who do we think we were, objecting to this and we're standing in the way of progress…

> The fact that we should really have an equal say in their policy decisions never crossed their minds (anti-fish farming group member).
>
> Our history of not getting answers from government is very bad, that is the biggest obstacle, we want to be working with departments and working in cooperation and it is not acceptable that government departments do not address our concerns and treat us with […] I don't want to use the word contempt […] but they just seem to ignore us really (medical environmental group member).
>
> Really and truly a lot of it is politics because, like for example county councillors, we found that in general they are very much involved with business interests, and it is amazing how little support you have when you go to them for reasons like ours that mightn't be as profitable or whatever (rural environmental watchdog group member).
>
> Our group funded and commissioned a study for an integrated coastal zone management strategy for [our area], sent copies to every conceivable politician, and got zero response. And the conclusion that we draw is that the Irish government is prepared to pay lip service to the environmental issues and to buy into protocols, in the full knowledge that they have no intention of upholding them. And we have documented every single communication that we've had. We have had a series, for example, of 18 different letters on the same subject with [government department name]. I've been promised five different times by five different people that within a subsequent week I would get answers to questions I posed. I'm still waiting for those. The last communication I had was a year ago, I gave up because I knew they would lie to me again. But that's what we're dealing with. We're dealing with people who are, on a daily basis, lying to us; that's our experience. They will tell us anything they possibly can so that we go away (local lake protection campaigner).

Some recent discussions about the importance of emotions in collective political protest (see, for example, Holmes [ed.] 2004, Goodwin *et al.*, 2001) argue that social movements recruit people who experience particular emotions – particularly anger, moral indignation and sentiments of injustice. Our study suggests that this is the wrong way round: anger at injustice and distrust of "authorities" are collective phenomena that arise out of group experiences and are mediated to individual members as they reflect over, discuss and mutually support each other through those experiences. Engagement in collective lines of action has a "radicalising" effect on participants. It generates a shared "sociocultural perspective on reality" (Fischer, 2005), combining a distinctive epistemology of the social with an attachment to particular values (justice, honesty, openness to learning) that are grounded in strong emotions. Out of a shared historical memory of experiences of rebuff, hostility and attempted manipulation, what Crossley (2002) calls "a resistance habitus" is formed, which leaves these individuals altered by the experience of environmental activism in a way that the personal environmentalists are not.

Activism, "greening" and practices of citizenship

Above I have presented two brief portraits of different ways of practising environmental citizenship. Here I try to draw out some implications of this. The "greened" citizens in our study, whom we labelled "personal" environmentalists, fit the concept of "civic privatism" if we redefine that concept somewhat: they are still "civic", in that they are trying to engage in their personal lives with environmental issues that confront them as citizens, but their mode of dealing with these is "privatised". Being a good citizen, in their understanding of this, is largely realised through the moralising and disciplining of the self, and they reject collective action because they fear that becoming a "serviceable agent" (or "cultural dupe") of the collectivity would require a compromising of the self – an abandonment of rational and reflexive monitoring of their own practices. Most of the "collectives" we encountered, on the other hand, reject the idea that joining a group means becoming its obedient servant; they consciously design themselves as "affinity groups" rather than as Social Movement Organisations, and they continue to negotiate their interactions with others in ways that allow them to avoid becoming labelled as simply a "group member". Nevertheless, it seems that belonging to a group does alter individual identity, as collective activists go through a process of collective learning, particularly about the political processes around environmental management in Ireland, which transforms their cognitive and emotional relationship to the state and to democracy as it operates in Ireland. The important contrast here is not between "passive" and "active" citizenship, but rather between two different ways of understanding the practice of citizenship.

As Leach *et al.* (2005) argue, environmental policy discourses are not just discourses about "the environment", but are also discourses that construct understandings of citizenship. The Irish state, and some of the more incorporated environmental NGOs, clearly prefer Irish citizens to be "greened" rather than "mobilised", as is evident, for example, in its strategies for managing the domestic waste crisis. Rather than addressing industrial and market generation of waste, the state has chosen to launch a very public campaign that targets the practices of the general public as the source of the problem, and urges Irish citizens individually to "reduce, reuse and recycle" to resolve it, while simultaneously increasing the cost of using the domestic bin service. While the state's openness to public consultation and "dialogue" around contentious planning issues has increased in the last decade, expert discourses (in contexts of opposition to plans for incinerators to deal with waste, to site designations under Natura 2000, to genetically modified foods or to road planning, and so on) still assume an unchallengeable authority for "sound science" and a "deficit" model of public knowledge. Citizens have a right to "participate" in decision-making, but to not recognise the superior knowledge of experts is irrational and can threaten the removal of participation rights.

This suggests that there has been a transition to a liberal-democratic conception of citizenship in Ireland, coherent with Leach *et al.*'s arguments about this model of citizenship. But it coexists ambiguously in Ireland with an older, "communitarian" view, in which a "developmental" or "modernising" state seeks to realise social and economic benefits for citizens by actively engaging them in a collective project (the project of modernity). The rhetoric of "community" and "nation" still pervades political and policy discourse, even as the state increasingly promotes a vision of that nation as "aggregated individuals/consumers". This often makes it quite difficult to "read off" the model of citizenship used by the state from its policy discourses.

"Modernisation" and individualisation, and particularly the transition to a risk society, are said to generate increasingly educated and critical publics who start to question the collective project and to prioritise individual rights. The experience of uncertainty, it is said, propels people into reflexivity. But in our study, the concept of the "reflexive citizen", as developed by Beck, DuPuis and others, turned out not to work as it was expected to. In our case, reflexive citizens emerged as "agenda-takers" rather than "agenda-setters" (Leach *et al.* [eds.], 2005) in relation to environmental problems. That is, they largely accept the diagnosis of these problems and their management, which emerges from expert discourse, and while often critical of individual politicians, they exhibit a generalised trust in the political system and in knowledge authorities as the basis on which they should strive to moralise their own practices. It is the collective environmentalists who question and distrust authorities and who push for "cognitive justice": acceptance of the equal worth and validity of citizen knowledge in relation to environmental issues. They strive to be agenda setters, although they are often able to achieve this only within civil society (and then only to a limited extent), not in the formal political sphere; over time this transforms into an obsession with holding those who govern them accountable for their actions.

The different understandings of citizenship and how to practise it, as shown here, could be summarised as the difference between "disciplined" and "performative" citizenship. Reflexive consumers can be seen as an example of "disciplined" citizenship, where discipline is exercised primarily over the self and as a way of realising a self-project as a moralised individual. The collective activists are examples of "performative citizens", where performance is directed outwards towards social critique and the realisation of a project of social change. Their goal is the democratisation of management of the environment, a project that develops within and as a result of collective mobilisation to protect nature – sometimes local nature ("place"), sometimes natural systems (habitats, resources). The disciplined citizens articulate a liberal vision of democracy, while for the performative citizens it is something that appears much closer to a civic republican model, in which the aim of democratic action is to overcome forms of domination and power over the individual, both that of the state and that of the collective group itself.

In conclusion, then, citizens respond to policy discourses and their implicit theories of citizenship in divergent ways. Some are formed and informed by a liberal perspective on citizenship; others challenge and seek to replace this with a civic republican perspective. While expert and state discourses play a role, a critical mediating factor is the personal experience of collective engagement. It remains important to differentiate between group and individualised engagement and to recognise the significance of collective dynamics in shaping practices of citizenship.

References

Bauman Z., *Globalisation: European Experiences*, New York, Columbia University Press, 1999.

Beck U., Beck-Gernsheim E., *Individualisation: Institutionalised Individualism and Its Social and Political Consequences*, London, Sage, 2002.

Boltanski L., Thévenot L., *De la justification: les économies de la grandeur*, Paris, Gallimard, 1991.

Buttel F., "The Recombinant BGH Controversy in the United States: towards a New Consumption Politics of Food", *Agriculture and Human Values*, 17, 1, 2000, pp. 5-20.

Crossley N., *Making Sense of Social Movements*, Buckingham, Open University Press, 2002.

Dubet F., Lustiger Thaler H., "Introduction: the Sociology of Social Action Reconsidered", *Current Sociology*, 52, 4 (Monograph 2), 2004, pp. 557-574.

DuPuis E., "Not in My Body: rBGH and the Rise of Organic Milk", *Agriculture and Human Values*, 17, 2000, pp. 285-295.

DuPuis E., *Nature's Perfect Food: How Milk Became America's Drink*, New York, New York University Press, 2002.

Eder K., "The Institutionalisation of Environmentalism: Ecological Discourse and the Second Transformation of the Public Sphere", in Lash S.,

Szerszynski B., Wynne B. (eds.), *Risk, Environment and Modernity*, London, Sage, 1996, pp. 203-223.

Farro A.L., "Actors, Conflicts and the Globalisation Movement", *Current Sociology*, 52, 4 (Monograph 2), 2004, pp. 633-648.

Fischer F., "Are Scientists Irrational? Risk Assessment in Practical Reason", in Leach M., Scoones I., Wynne B. (eds.), *Science and Citizens: Globalisation and the Challenge of Engagement*, London, Zed Books, 2005, pp. 54-65.

Giddens A., *Modernity and Self-Identity: Self and Society in the Late Modern Age*, Stanford, Stanford University Press, 1991.

Goodman D., "Agro-food Studies in the 'Age of Ecology': Nature, Corporeality, Bio-politics", *Sociologia Ruralis*, 39, 1, 1999, pp. 17-38.

Goodman D., DuPuis E.M., "Knowing Food and Growing Food: beyond the Production-Consumption Debate in the Sociology Agriculture", *Sociologia Ruralis*, 42, 1, 2002, pp. 6-23.

Goodwin J., Jasper J.M., Poletta F., *Passionate Politics: Emotions and Social Movements*, Chicago, University of Chicago Press, 2001.

Holmes M. (ed.), "Special Issue on Anger in Social Life", *European Journal of Social Theory*, 7, 2, 2004.

Jamison A., *The Making of Green Knowledge – Environmental Politics and Cultural Transformation*, Cambridge, Cambridge University Press, 2001.

Leach M., Scoones I., Wynne B. (eds.), *Science and Citizens: Globalisation and the Challenge of Engagement*, London, Zed Books, 2005.

Lichterman P., *The Search for Political Community: American Activists Reinventing Commitment*, Cambridge, Cambridge University Press, 1996.

McDonald K., "Oneself as Another: from Social Movement to Experience Movement", *Current Sociology*, 52, 4 (Monograph 2), 2004, pp. 575-594.

Spaargaren G., Mol A., Buttel F. (eds), *Environment and Global Modernity*, London, Sage, 2000.

Tovey H., *Environmentalism in Ireland: Movement and Activists*, Dublin, Institute of Public Administration, 2007.

Wright W., Middendorf G. (eds), *The Fight Over Food: Producers, Consumers and Activists Challenge the Global Food System*, University Park, Pa, Pennsylvania State University Press, 2008.

Mistaking Publics

A Challenge for Environmental Governance

Luigi PELLIZZONI

Associate Professor, University of Trieste, Italia

In this chapter I wish to reflect on recent trends in environmental governance, focusing on a type of paradox. New policy instruments appear to be creating stronger links between policy makers and addressees, policy formulation and implementation, strengthening the role of the public. Yet looking closer, their rationale seems at odds with such a goal or outcome. Regulatory arrangements may mistake the public they play a role in building, failing to address the latter's actual interests and concerns.

This argument will be developed through exploring the notion of public, looking at the conditions of use of environmental resources, examining the controversy over the latest generation of governance arrangements, elaborating on the latter's logic and briefly reflecting on alternative directions.

The problem of the public

Why is it so difficult to protect the environment? One possible reply is uncertainty. Perhaps we do not know enough about it. Right or wrong, this argument lends itself to controversial use. Even widely recognised environmental problems such as climate change are played down by powerful lobbies that build their sceptical arguments on an alleged lack of sufficient scientific evidence (Freudenburg *et al.*, 2008; Jacques *et al.*, 2008). And should precaution be seen as a key to pre-empting major disasters, or a flawed idea leading to "financial losses, restricted freedoms and the foregone health and environmental benefits of restricted technologies" (Wiener and Rogers, 2002: 321)? Undoubtedly, there are contrasting preferences on risk regulation (Pellizzoni, 2009). Industry and scientists are more concerned with reducing false positives (evidence of harm turning out to be incorrect), because the latter directly affect the profitability of investments and the continuation of research. Technology end users and hazardous plant neighbours are, on the contrary, more concerned with reducing false negatives (no evidence of harm turning out to be incorrect), because the latter directly affect their health and well-being. The knowledge deemed sufficient to decide on policy measures is related to the heatedness of conflicts: the deeper the latter, the stronger the evidence required (Sarewitz, 2004).

Behind controversies over knowledge, therefore, we often find controversies over the distribution of environmental positives and negatives. As a consequence, the notion of externality is no less important than the notion of uncertainty. Transactions may produce effects on agents who did not take part or have a say in them. Externalities, like pollution, are often negative. Negative externalities indicate ineffective resource allocation and unacknowledged interdependencies. From this viewpoint, answering ecological questions means – as economists insist – internalising externalities and reorganising transactions so as to consider previously neglected interdependencies.

Acknowledging and internalising interdependencies is what John Dewey describes as the formation of a public. Compare his idea of the public with mainstream ones: Charles Taylor's, for example. According to the latter (Taylor, 2004), the notion of the public builds on two semantic fields: matters of communal interest (public affairs, public authority, etc.) and matters of access and expression (public availability, public news, etc.). Though supported by a considerable literature (Weintraub, 1997), this view is problematic. For example, according to it, economic transactions are private in both senses: they do not call for communal decisions, nor do third parties have access to them. Yet what about externalities? What about the public role that, ever since Adam Smith's invisible hand, the indirect consequences of economic transactions have been acknowledged to play?

Taylor takes "community" as an unquestioned reference point, which is often not the case, and considers two codes of public discourse (Ku, 2000): inclusion-exclusion (who talks) and openness-secrecy (what do we talk about). Yet there is a third one: accountability-unaccountability (how do we talk about it). This is what Dewey focuses on: "The public consists of all those who are affected by the indirect consequences of transactions to such an extent that it is deemed necessary to have those consequences systematically cared for" (1927: 245-6). Transactions, therefore, are public if their consequences (are deemed to) spread beyond the directly involved actors. The essence of publicness lies in the acknowledgement of problematic interdependencies, the existence of third parties affected by our transactions to whom we make ourselves answerable. As a consequence, evaluative or justificatory criteria transcend the participants in transactions, including forms of external accountability.

Of course, the threshold between private and public is questionable, depending on awareness of the possible consequences of actions and how a consequence is defined; it also depends on normative judgements about the consequences to be controlled and about who is regarded as being directly and indirectly involved (Geuss, 2003). The very notion of third party is tricky. A public is composed of subjects lying outside a given relational setting, or seen in a different perspective, equipped with a novel identity. They are, or become, to some extent "strangers" to us (Gurevitch, 1988). Yet they cannot be totally extraneous, otherwise there would be no shared element enabling comparison (Lyotard, 1983). Thus, the third is not "one of us" but at the same time belongs to a broader "us" that we grasp but need to specify. The search for the public offers only provisional, tentative endpoints.

Similarly, Boltanski and Thévenot (1991; 1999) distinguish between public and private contexts of interaction according to whether or not a reference is made to general, impersonal criteria. The latter vary according to what is defined as worthy, yet in a public context a course of action always appeals to an idea of community, of common good, of world in common. Allocation of material and immaterial resources is justified so long as it respects, at the same time, equality and hierarchy, common belongings or interests and value rankings deemed appropriate to the chosen regime of worth. So, for example, in what Boltanski and Thévenot call the "domestic regime", environmental protection is connected with traditions, local specificities, territorial vocations (with related rigidities and privileges), while in the "industrial regime" what is stressed is the centrality of technical and economic efficiency and its champions (experts, entrepreneurs).

A regime of worth, therefore, establishes a comprehensive division of labour, a solidarity system in a Durkheimian sense, with no residual unaddressed claims or unwarranted imbalances. There is, as a consequence, a constant tension between recognition and disappearance of the public. Whatever the criteria adopted, when all interdependencies are allegedly recognised and internalised, the distinction between internal and external blurs and the public disappears. What remains is only "us", with no further distinctions. We talk among us and about us. Public discourses become private ones. Justifications drift towards self-reference. Broadening the notion of the public to purportedly encompass the whole community thus amounts to negating it. The public appears and survives only as otherness, dissonant discourses or neglected interdependencies, an unexpected turn in the course of events.

Goods and publics

A well-known classification of the conditions of use of goods (Ostrom *et al.*, 1994) suggests that the latter can be distinguished according to their excludability (the possibility of preventing others' use) and subtractability or rivalry (the extent to which one's use affects others'). Public goods are characterised by difficult excludability and low rivalry. Access is free and non-competitive. As a consequence there is in principle no need (or possibility) of regulation.[1] Such need arises only when subtraction becomes a relevant issue. Private goods are characterised by easy subtractability and excludability, thanks to property rights, for example. Club or toll goods[2] are characterised by difficult subtractability and easy excludability, thanks to legal or technical barriers, for example. Think of theatres, toll roads, private schools or golf clubs: only those entitled can enjoy the good, yet their use remains reciprocally unaffected. Finally, commons are characterised by easy subtractability and difficult excludability. Think of open rivers or grazing lands: users have equal access and competing interests in using such resources, without being compelled to take care of their maintenance.

What is important for us here is that different conditions of use lead to different configurations of externalities, or publics (Figure 1). In ideal-typical terms, we can start by imagining a community provided with plentiful goods, freely available to everyone. Emerging scarcity transforms some of these goods into commons. The very social or physical factors responsible for this will inevitably advantage some to the detriment of others. The originally undifferentiated community splits into groups of insiders and outsiders. The former have direct access to a resource; the latter may of course try to get inside. The situation is especially difficult if the insiders' use is rival and it is complicated to build collective barriers to entrance. This may incite them to parcel out the goods and establish property rights. The situation is more relaxed if the insiders' use is non-rival and collective barriers can be raised with relative ease to keep out other groups. Growing scarcity or difficulty in maintaining the barriers may lead, however, to shareholder rivalry or strengthened external pressure. Again, this may incite them to parcel out the goods.

1 Obviously this does not always apply to the production of such goods. While some, like sunshine, do not require any social contribution in this respect, others, like the army or the police service, do need such contribution, which has to be regulated.

2 Toll and club goods are usually treated as synonyms in the related literature, and here I shall stick to this practice. A difference, however, may be traced (Prakash, 2000): toll goods are those whose use can be unitised (and priced) while club goods are those for which this is difficult or impossible. The price paid in the first case for each single unit of the good is replaced in the second case by a membership fee.

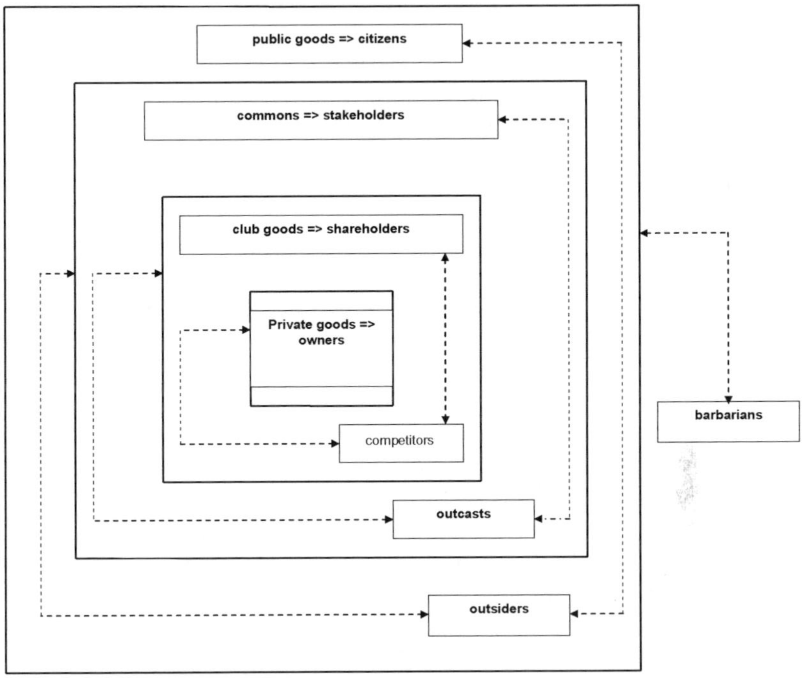

Figure 1 – Conditions of use and publics

The figure can also be read the other way round, or without any fixed order. When goods are private we can distinguish between an inner constituency of owners or authorised users, whose interests have been satisfied through negotiation, and an outer constituency of affected non-users who take the form of carriers of unaccommodated, competing interests in the goods. Members of a club, on the other hand, share non-rival interests in their goods. Protected by technical or legal boundaries, these shareholders see the outer constituency of affected non-users as composed of outcasts, rather than true competitors. In the case of commons, the inner constituency is composed of users with free access but competing interests. Beyond this circle of stakeholders lies a public of outsiders who, should they wish to enter, would not have tolls of any relevance to pay. Finally, users of public goods regard each other as fellow citizens, members of a comprehensive constituency. They have equal, non-competitive access to the resource. They recognise no third party because they see no interests and concerns other than their own.

Yet, as remarked, the equation public=community-as-a-whole entails circularity. This is actually borne out by the discursive procedures applied to public goods. The expression itself is in a sense paradoxical.

If a public is created by the externalities of the use of a good, then public goods lack a public.[3] In any condition of use the existence of a common origin, a shared background encompassing "us" and "them" in a broader "us", is always premised on tracking down externalities. Yet if, as with sunshine or national security, "everyone" is included, there are no externalities. In similar cases, justifications seeking to legitimise the availability of a good (human-sourced or otherwise) typically appeal to some transcendent point of reference (God, Nature, the People, the State, Justice, etc.) – a fictitious third party hiding a self-referential procedure. Of course, in most cases a proper public does indeed exist, yet the appeal to a general will, interest or principle obscures it. Outside a comprehensively defined community one can find nothing – beyond irrelevant commitments or incomprehensible claims, like those of barbarians for the citizens of Greek poleis.[4] Questioning a regime of worth, and bringing to the forefront neglected interdependencies, in fact means claiming the existence of involved parties lying beyond acknowledged constituencies, established boundaries of worth.

In this framework the current issue of environmental governance can be described as follows. Conditions of use of the majority of goods are not stable but change, in most cases according to social factors[5] (techno-science, demography, economics, politics, law, culture). A growing variety of goods previously unaffected by use are taking on the character of commons because of the increased number of users and their growing ability to exploit such resources.[6] Commons, however, suffer from over-consumption (the "tragedy" famously described by Garrett Hardin). Expanding regulation is thus mandatory. Neither-state-nor-market institutions for managing commons have worked for centuries by carefully distributing burdens, privileges, controls and sanctioning powers. Yet the conditions for success of such institutions are not easily met for

---

3 This is obviously because in the expression "public goods" the term "public" is used to mean "pertaining to all of us" rather than "pertaining (also) to third parties". It would indeed be sensible to find another label for this type of goods; for example, "stable goods".

4 It may be worth a reminder that in its original meaning – to be found in Plato or Aristotle – the term "barbarians" refers to people who speak a totally different language, so it is impossible to dialogue with them, to grasp and consider their claims.

5 There are modifications that cannot reasonably be ascribed to human action; for example, climate change in the pre-industrial era.

6 Indeed, even air does not appear to be a public good any more – at least not to city dwellers competing for it with cars and factories.

many commons in today's society.[7] This seems to restrict policy options to two alternatives: either state authority or private property.

To the extent that it impinges on the conditions of use of resources, regulation also affects the public, the outer constituency of involved non-users. Assessing the impact of environmental policies, thus, means answering the question: what is the public they play a role in building and to what extent are its interests and concerns cared for? Mistaking the public means mistakenly or misleadingly addressing its concerns. In particular, blurring the distinction between the good of the many and the good of the few is always a pitfall, because policies can claim to be "public" only to the extent that they make a case for their ability to accommodate special and general interests, to internalise externalities or to keep only those whose benefits outweigh their costs (for the whole of the society or even for those who bear the costs).

Environmental governance

The relationship between private and public domains may therefore offer a clue to the rationale and implications of current trends in environmental governance. The latter's pace of innovation has led some people to talk of a "silent revolution" (Theys, 2000) – from government to governance, from state-centred steering of public affairs to formal and informal interaction, partnership and cooperation between public and private actors, or the self-regulation of the latter.

Be it the deliberate outcome of neo-liberal reforms or an unforeseen effect of the state crisis vis-à-vis techno-scientific advancement and economic globalisation (Strange, 1996), governance in the environmental field largely corresponds with the emergence of "third generation" approaches: command-and-control regulation (targets such as emission limits for pollutants and penalties to be applied if such targets are not met) and market-based regulation (taxes, incentives, tradable permits to pollute). This generation includes three main categories of instruments that variously apply open negotiation and self-regulation (Prakash and Kollman, 2004): mandatory information disclosure through labels or emission registers such as the US Toxic Release Inventory Program; business-government partnerships such as the US 33/50 and Project XL programmes, or the Dutch covenants; and government and non-government-sourced management systems (standards, codes of conduct, certi-

7 For example, the features of the resource and the effects of its use should be well known; users should belong to a network of established relationships and use the resource in similar ways; legal rights, traditions or physical boundaries should act as limits to widespread access (Ostrom *et al.*, 1994). On this point see also the recent collection of essays included in Baland *et al.* (2007).

fications) such as ISO 14001, EMAS, the US chemical industry's Responsible Care Program or the certification system of the Forest Stewardship Council (FSC).[8]

These instruments can be regarded as part of a broader family of corporate social responsibility (CSR) approaches aimed at contributing to sustainable development and enhancing quality of life, the common feature of which is to be found in their voluntariness (Bendell and Kearins, 2005). In its turn, CSR is part of the growth of "civil regulation" (Vogel, 2006) or "private governments" that include, for example, the so-called *lex mercatoria*, the corpus of trade usages developed outside national legislation (Teubner, 2002). Another side of civil regulation can be seen in critical consumerism. Consumers increasingly choose "producers and products with the goal of changing objectionable institutional or market practices" (Micheletti *et al.*, 2004: xiv), according to considerations of justice, fairness, personal and family well-being, animal welfare, or environmental protection. Through boycotts, "buycotts" (selective shopping) and symbolic attacks on producer or product images ("culture jamming", "subvertising", etc.) consumers perform ethical or political assessments of business and government practices. Critical consumerism thus operates via market relationships but according to a self-regulatory standpoint that goes beyond the logic of money.

Instead of a stable representation of interests, as with neo-corporatist arrangements, we are therefore increasingly faced with ad hoc coalitions in a problem-solving framework. According to their supporters (Prakash and Kollman, 2004), third-generation environmental policies and, more generally, CSR initiatives and private governments effectively address the problems of command-and-control and market-based regulation: the two straightforward answers to the tragedy of the commons. The former allegedly leads to over-regulation; it is unable to follow the dynamics of technological and economic transformation and to adapt to specific social or environmental conditions; it suffers from knowledge gaps regarding the environmental and health impacts of human activities; it may lead to spending money on relatively insignificant risks; it requires effective and costly monitoring and sanctioning systems. Market-based instruments are more flexible yet they also require robust knowledge of technical and financial conditions (for example, to be effective a tax should be neither too heavy nor too light), adequate monitoring and sanctioning capacity and well-specified property rights – that is, state-

8 I have chosen here what seems to me the simplest classification available. There are others. For example, some people regard mandatory information disclosure as a kind of market-based incentive (Konan and Cohen, 1997). Controversial classification is an indicator of the novelty of these instruments.

centred institutions and regulations. Moreover, tradable permits legitimise arbitrarily settled levels of pollution and may promote relocation of polluting activities to less expensive neighbourhoods inhabited by disadvantaged groups.

Third generation approaches, on the other hand, build on the assumption that the best way to regulate is to rely on the insight and motivation of the regulated. The assumption, in other words, is that public goods such as the protection of the environment and human health are better ensured by strengthening the role of private means, by promoting "beyond compliance" corporate behaviour and building on the direct interaction between public and private actors, or simply between private actors. In this sense, compared to market-based instruments, third generation instruments take a further step away from the "state" solution to the tragedy of the commons – and a more sophisticated one, since they do not simply transform environmental resources into marketable goods. Moreover, by fostering horizontal networks and corporate or consumer direct assumption of collective responsibilities, they point to flexible ways of assessing and harmonising private and public interests, expanding participation and widening inclusion in the policy processes, as testified by the amazing increase in controls (Power, 1997) through third party certification, stakeholder involvement or communication with public opinion at large.

Of course, command-and-control and market-based regulation are not withdrawn, but they mesh in various ways with third generation regulation. For example, information disclosure requirements are no longer mandatory in the sense of command-and-control regulation (Van den Burgh and Mol, 2008). Firms remain free to choose their targets, on the assumption that they are interested in offering a "green" image to customers, suppliers, bondholders and other non-contractual stakeholders (from neighbouring communities to critical consumers). On the other hand, improving compliance with regulations and international standards is a major reason for the adoption of voluntary programmes such as EMAS and ISO 14001, or at least a major result of this (Kollman and Prakash, 2001; Falkner, 2003; Prakash, 2005). Governance also strengthens through an expansion of traditional market-based instruments. An example comes from the extension of the patent system to the biotechnology field. In this puzzling array and interconnection of solutions, the only common trait appears to be their basic reliance on a contractual logic, through the increase in property rights, formal deals or single-handed commitments towards specified or unspecified "counterparts" (from product end users to disadvantaged workers, from endangered species to fellow citizens).

Mistaking publics

Widely praised as the embodiment of "smart regulation" (Gunningham and Grabosky, 1998), third-generation environmental instruments remain, however, controversial (Steinzor, 1998; Kollman and Prakash, 2001; Prakash, 2005). Critics remark, for example, that corporate ecological commitments may consist of mere "greenwashing" (Laufer, 2003). Sectoral targets may encourage free riding, whereby polluters take advantage of the improved performance of other firms (Börkey and Lévêque, 2000). Agreements may be used to postpone or forgo stricter command-and-control regulation (EC, 1997). Independent verifiers may suffer from interest capture (Kerwer, 2008). Moreover, information asymmetries and legal risks involved in certifications may lead verifiers to concentrate on documents rather than facts, formal requirements rather than substantive outcomes (Power, 1997; Kimerling, 2001).

Admittedly, comparing voluntary and self-regulatory approaches with command-and-control and market-based regulation is not easy (EEA, 1997). Some use of counterfactual reasoning (what if…?) is inevitable. Moreover, even when appropriate, criticisms do not necessarily point to any inherent flaw. They may just show that new policy instruments are in need of tuning. Yet perhaps there is more to the picture than meets the eye, as the perspective developed in the preceding pages may help to clarify.

While a growing number of environmental resources take the character of commons, third generation approaches eschew state control without turning to straightforward privatisations. Their basic strategy is to create club-like arrangements (Figure 2). Think of business-government agreements, environmental management schemes, codes of conduct, labels and emission registers. Adhering firms share a non-rival interest in the maintenance of such partnerships or systems. Those who adopt the EMAS, for example, bear the related costs because they draw economic advantages from the efficiency and credibility of the scheme. Individual benefits can merge in this way with collective ones, and the public of interested non-users of the system may gain insight into its workings through forms of external accountability, such as environmental reports and independent certifications.

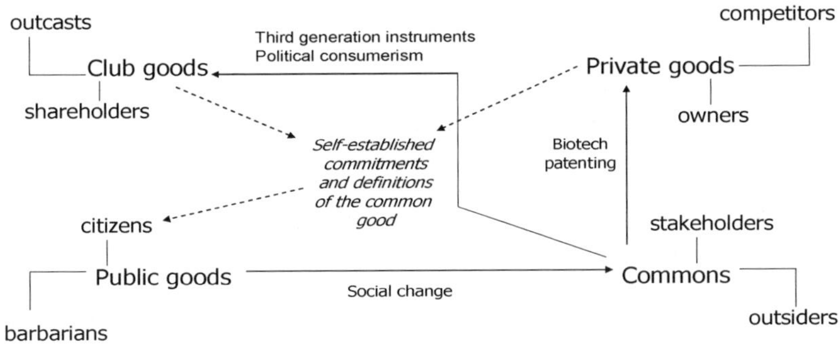

Figure 2 – Mistaking publics

However, there is a difference between "advocacy for humans and effective representation of human interests in very technical and legally obfuscated initiatives" (NGO member, quoted in Steinzor, 1998: 125). Whatever particular interdependencies are considered, and however the balance of public and private interests is achieved, it is largely in the hands of club members. State-of-the-art environmental performance, for example, is not necessarily a target, nor is it easy to assess. Verification usually focuses on whether and how a goal has been achieved or pursued, rather than why such a goal has been set (Harrison, 1999; Cashore *et al.*, 2004). This is not a matter of weak design or implementation, but of the very logic of voluntary and self-regulation. The alleged efficiency of the latter vis-à-vis growing difficulties in setting reliable, effective output standards depends precisely on the fact that goals and means are negotiated or defined by the firms themselves, drawing on their privileged insight into their own business and task environment. Stakeholders and auditors have few means of assessing whether more ambitious targets have been discarded because of cost, technical trickiness or market implications.

Accountable actors, in other words, set the frame for their accountability. They can be made answerable for nothing more than their voluntarily assumed commitments. No overarching rule can be invoked to define public interests and their balance with private ones. In this sense, as theoretically anticipated, club-like settings produce outcast publics. This is not a problem of environmental regulation only. Students of CSR remark that "partnerships can develop only where the company is interested in achieving the goal concerned… The range and level of obligations [firms] are expected to fulfil are largely left to their discretion" (Newell, 2005: 545-546), while "crucial economic issues tend to be excluded from the contents of CSR standards" (Frynas, 2005: 587), for example, as regards the impact of industrial infrastructure on the subsistence of local people, firms' freedom to invest and divest at will, and the drawbacks for national economies of a heavy reliance on the export of natural resources.

Major criticisms levelled at third generation approaches thus focus on the way public interests are adapted to private ones. Club-like regulatory arrangements produce outcast publics because corporate commitments as regards questions, goals and means are self-defined, and the terms of inclusion of public concerns are self-established. Accounting schemes and information disclosure provide external insight, yet not necessarily into what outcast constituencies regard as most urgent and relevant. Allegedly shared interests and concerns easily hide or produce conflicting ones. Recognising such conflicts would amount to unveiling a public of competitors, with the probably consequent plea for adequate means to defend their interests. This, however, is hampered by the very crafting of the public as consisting of outcast parties, placed under "green" guardianship. Compared with traditional privatisation policies, third generation ones may therefore make it harder to reveal the specific interests lying behind a cover of alleged general interests, to discern how real publics have been blurred to create a supposedly comprehensive public.

In other words, the inherent weakness or contradiction of third generation instruments is that they rely on strengthened accountability yet develop major deficits precisely in this respect. They evoke an imaginary encompassing public while producing outcast parties entrusted to corporate or professional care, thus provided with no or very little room for counteraction, as would likely be the case if they were recognised as proper competitors. Of course, suitable designs can address this problem. For example, FSC seeks to balance out the decision-making power of social, environmental and economic interests, with some success (Gulbrandsen, 2008). However, greater inclusiveness does not necessarily solve the problem. It may just broaden the circle of self-referring actors. As a consequence, residual externalities may become more difficult to recognise (Pellizzoni, 2008). Remember that, according to Dewey, publicness does not depend on who is included, but on how the affected outer constituencies are dealt with. Thus, for example, stakeholder representatives are often bound to accept the issue-framing established by the accountable actor; they may be captured by answerable interests; they may defend only their own interests, taking stances that do not necessarily coincide with unrepresented concerns (Bendell and Kearins, 2005).

A deficit of accountability also surfaces in critical consumerism. The latter rests on the assumption that "citizen concern for their private lives can be used in a beneficial way for society at large; [that] privately oriented virtues have a public role to play" (Micheletti, 2003: 159-160). Such concern may lead to traditional forms of mobilisation (demonstrations, lobbying, etc.). Yet while in most mobilisations participants seek to gain something for themselves, in this case the outcomes sought are beneficial to other human or non-human beings. Critical consumers thus behave as club members who share a non-rival interest (in getting consumption "right") and see third parties as an outcast public under their own, or corporate, guardianship. Yet unaccounted-for, competing interests may surface in this case as well. Consider the side effects of boycotts aimed at modifying working conditions in developing countries: restricting young-worker employment, for example, may lead to shrinking family income (Vogel 2006). Protecting third-party interests becomes a less straightforward endeavour than expected.

Moreover, even if, as many scholars contend, critical consumerism is primarily "individualised collective action" (Micheletti, 2003) – its political impact stemming from the aggregate effect of individual shopping choices – it ultimately relies on personal specifications of firms' answerability, own assessments of the public good and the way the latter is affected by the individual's buying behaviour. So again, the distinction between, and harmonisation of, public and private interests turn out to be approached in a solipsistic way.[9] According to the famous definition of Carl Schmitt (1922), sovereignty consists of the capacity to decide upon the exception to the rule; that is, to redefine the existing order, to establish a new one. In this sense consumers are really "sovereigns", as usually depicted. Namely, they are sovereigns not only of their private choices of taste but of the distinction between private and public concerns. No overriding rule can be applied to define and balance them. Consumers are rulers of themselves.

Something similar happens with biotechnology patents. This case is interesting because it shows that mistaken or misled publics derive not only from new regulatory arrangements, but also from more straightforward ways of privatising commons. New biotechnologies act at the level of cells or genes. The latter thus become sensitive to use.[10] They can no longer be regarded as public goods but as commons. Competing interests emerge and require regulation, which has taken the form of an extension of the patent system: an expansion of property rights. To

9 Research indicates some problematic consequences. For example, critical consumers seem more likely to be agenda takers than agenda setters, accepting government or corporate issue-framings more easily than organised groups (Tovey, 2005).

10 Compared with new biotechnologies, old forms of manipulation of human genetic patrimonies such as war, racial segregation or genocide look exceptional and rough rather than "ordinary" and subtle.

include living beings in the realm of artefacts is claimed to be in the public interest, because innovation is regarded as per se beneficial to society and because, by ensuring that knowledge is revealed and economically viable, patents are regarded as the best way to promote and spread innovation. Both assumptions are obviously debatable. The latter, in particular, has been brought into question by the "tragedy of the anti-commons" (Heller and Eisenberg, 1998). Multiplying intellectual property rights and commercial interests may actually hamper innovation. An example is offered by the patents on so-called ESTs (Expressed Sequence Tags). Since the 1990s US patents have been awarded on genetic fragments with no corresponding gene or protein, biological function or possible commercial application having been singled out. Such patents entail complex and expensive transactions, since both scientific research and the development of commercial products require access to many patented fragments. The net result is an increase in costs and a slowdown of research (Rai and Eisenberg, 2003).

This drawback is important because it impinges on an assumption regarding patents in general. Yet the extension of the patent system to living beings also draws on specific assumptions: that for the sake of innovation, access to "necessary raw materials"[11] of any type is to be allowed, including human biological matter; and that such access is to be open "to those who have *legitimate* research interests in their use and *presumably possess the capability* to perform sophisticated scientific studies that can reveal biological information about the samples or even health-related information about the persons from whom they came" (NBAC, 1999: 59, italics added).

In this way specific actors are identified who, by virtue of their role – as testified by formal records in science and business, and validated by public authorities such as patenting offices and the courts – are entitled to enter the most intimate sphere of life, including human life. More precisely, by virtue of their validated role these actors are entitled – case by case, issue by issue – to decide upon respect of individual privacy and access to functional capacities. They are also entitled to decide what is to be regarded as proprietary to single human beings and what is to be regarded as proprietary to agents allegedly working for the common

---

11 Cf. the following statement: "Yet one earnestly wish to protect privacy and dignity without accepting […] that the interference with those interests amounts to a conversion of personal property. Nor is it necessary to force the round pegs of 'privacy' and 'dignity' into the square hole of 'property' in order to protect the patients, since the fiduciary-duty and informed-consent theory protect these interests directly by requiring full disclosure. […] The extension of conversion law into this area will hinder research by restricting access to the necessary raw materials" (US Supreme Court, Moore v. Regents of University of California, 51 Cal. 3d [1990] at 140: 144-145).

good. They are therefore mandated to define where the common good lies in the specific case.[12]

Contracting worlds

I have already said that if a common trait is to be found in the latest wave of environmental policies, it lies in the generalised adoption of a contractual logic. Can the problems outlined be traced back to this logic?

Consider the basic structure of a contract. Actors are understood as rational, autonomous, sovereign individuals (*i.e.* undivided, self-sustaining units). Individuals are only contingently tied and definitively released after completion of the exchange. Contracts free them from undesired social bonds, allowing them to fulfil their own desires without engaging in personal relationships with others (Godbout, 1992). Contract-based regulation, therefore, has on its side not only a promise of efficiency but also consistency with core modern values: freedom, subjectivity, individuality and immunisation from communal constraints (Esposito, 2002; 2008). In contracts, moreover, power asymmetries disappear behind the formal equivalence of counterparts, which implies perfect symmetry of exchange, whatever the character of the goods. In addition, no other concern is relevant unless specified and accepted by the contracting parties. The latter are, by definition, stake and goal setters. To be considered, third-party interests must therefore be made to fit in with the transaction. As a result, the public is tailor-made, defined according to the framework of the deal.

Systems theory offers a neat account of this. For Luhmann "in a fully individualised, functionally differentiated society, any individual system can perceive external inputs only in terms of 'perturbations' or 'irritations' that to become meaningful need to be interpreted according to its own code" (Luhmann, 1993: 494).[13] That is why political steering is becoming increasingly problematic. If politics operates through the exercise of power, and the economy is sensitive only to money, the former cannot drive the latter but only aspire to promote its self-steering, a self-amendment in the desired direction (Luhmann, 1997). As

12 This entitlement is especially sanctioned by an inversion of the burden of proof that, after a report of the US Office of Technology Assessment (OTA, 1987), has entered the courts as an unquestioned standpoint: manufacture is presumed by virtue of the very demand of a patent; it is therefore the denying authority that must prove that this is not the case, *i.e.* that something "exists in nature".

13 In Luhmann's theory, codes are binary oppositions by which systems elaborate information from the environment, producing their own elements of meaning. Codes, in other words, allow the self-reproduction of systems. For example, science applies the true/false code, whereas law applies the right/wrong code.

citizens, people may be committed to reducing environmental damage or fostering development. As entrepreneurs, they cannot but look at the cost effectiveness of their business – if they are to survive.

How do such "irritations" work? A proper translation is impossible. Each system works according to its own code. There is 'an irreconcilable cleavage" between language games; one language game does not and cannot "exchange" elements with another one. A language game can only be provoked to link up with a sentence that makes part of the other language game. No exchange takes place; rather it is a "re-enactment […] [which] is neither translation nor trans-substantiation of the old element, but an independent reaction to something else by which the game creates a new element. […] Thus, a language game never 'contains' elements of another game, but only its own elements that 'link up' to elements of the other game" (Teubner, 2002: 207).

This is what is called a "re-entry". "A distinction re-enters itself if it is copied into itself. It then reappears as part of its own space, as part of what it distinguishes" (Luhmann, 1993: 485).

> Whenever we make an 'observation' we draw a 'distinction' of two sides and make an 'indication' of one of them. […] Now the distinction between the two sides makes a 're-entry' into one of these two sides; it reappears in itself. […] Then it is no longer the old distinction. It is the 'representation' of the distinction within one of its poles. It is the 'internalisation' of the external/internal distinction (Teubner, 2002: 205).

Re-entry thus designates a process by which an observation – the distinction of something from something else (an act of sovereignty in Schmitt's terms) – is reproduced within one of the poles of the distinction. The public becomes, in this way, a distinction internal to the private pole of the distinction between private and public. What is public and what is private is privately established. The external becomes a category of the internal, with no more substance than what contracting parties confer upon it. The third party is included only in the sense that it re-enters as a codified description. The differentiation between inner and outer sides is internalised and so becomes "visible" and "meaningful". This happens, for example, when cost-benefit analyses or insurance programmes re-enter the difference between monetary and non-monetary values by fixing a monetary value to the latter, as with the loss of a human life; when firms re-enter the distinction between profit and environmental protection or community development by assessing the profitability of ecological or development programmes; and when consumers re-enter the difference between tastes and ethical or political issues in their own buying behaviours, that is through choices of taste.

So, if there is no public space, as such, and no possible merging of "colliding discourses" (Teubner, 1996), what are the results of a "link-up" that operates through "re-entries"? If a tie between functionally differentiated spheres is still needed – if society is to survive – conceiving of something intrinsically extraneous to the modern logic of separation and immunisation is hard, which is confirmed by systems theorists' awkward choice of terms: "resonance" (Luhmann, 1986); "structural coupling" (Luhmann, 1993; Teubner, 2002); "transjunctional operations" (Teubner, 1997); and "productive misunderstandings" (Teubner, 2002). Yet if Luhmann is ambivalent on this point, Teubner is more explicit. He maintains that the spread of autonomous private governments is capable of effectively replacing the old social order, precisely by means of such misunderstandings. According to this view, publicness expands not despite but because of the fact that the relationship with third parties is misleading. Firms, for example, address consumers' ethical or ecological questions because they interpret the latter as economic issues (demand shifts), the only ones that make sense for them, while consumers welcome the consequent changes in production because they mistake them for ethical or ecological answers. In developing countries, firms read community needs as opportunities for philanthropic gestures to "calm them down", while local people interpret their own deprivation as an entitlement to receive gifts (Frynas, 2005). It could be argued that consumers or local people understand firms perfectly and do not expect any sincere commitment from them. Yet consumer and community concerns are *not* just economic while, according to this view, those of firms are only economic. A basic mismatch of meanings remains, and the bet is that non-economic aims can be re-entered in full as economic goals.

More generally, what ensures that any desired outcome will be achieved? Why should misunderstandings be productive rather than destructive? Who decides – and on what basis – that an outcome is positive or negative, and for whom? The very notion of productive misunderstanding is contradictory, because a process with no drivers can have no purpose. Governance seems to work as a new version of the invisible hand: a quasi-magical meshing of fully independent and reciprocally insensitive spheres of action. Such reformulation, however, is even more problematic than the original. For Adam Smith, the invisible hand rested on a robust network of social ties (Sen, 1987), now remarkably weakened. Moreover, it was supposed to work, as it were, "automatically": individuals contributed to the common good by simply looking after their own interests. In the context of today's governance, they now decide on both their own good and the common good. Win-win outcomes, however, can hardly be taken for granted. Systems interaction in full-size, real life situations, is highly unpredictable – let alone driveable (Funtowicz and Ravetz, 1999). Ethical consumers actually foster new market segments, leading to increased resource depletion and waste production, while land in developing countries is dotted with "non-functioning white elephants" (Frynas, 2005: 587) – unfinished buildings, unused machines, broken devices: a material testimony to dialogues of the deaf, misunderstood misunderstandings between companies or developing agencies and local people.

Conclusion

In this chapter I have focused on a paradox of environmental governance. Regulation expands as a consequence of social and technical change. Labels and registers, open negotiations, accounts and independent controls reframe or replace command-and-control and market-based approaches. Contract-based regulation seems to improve policy efficiency and individual responsibility-taking. Yet by adopting the self-referential logic of the contract, regulators are led to take their mirror reflections for real publics. The more encompassing the accounts, the lesser the probability that they will be able to grasp actual externalities and proper otherness. From this viewpoint, the often stressed increase in policy reflexivity (Fiorino, 1999) takes on a problematic connotation.

Third generation policy approaches typically build club-like arrangements, with outcast parties – mistaken for comprehensive publics – replacing competing ones. Yet in many cases competition does not disappear, it is just obscured. Third party positions are mistaken and addressed in a misleading way. As a result, exclusion may be broadened rather than restricted as contended by supporters of these instruments – an outcome further strengthened by the extension of straightforward privatisation policies, such as biotechnology patents.

The point, of course, is not to reject third generation policies in their entirety. A revival of pure command-and-control regulation would hardly be desirable or effective in a globalised society. In addition, the literature reports cases of successful dialogue and constructive relationships of accountability in a variety of contexts, ranging from Swedish eco-labelling (Boström, 2006) to Nigerian community development initiatives (Frynas, 2005). Yet, to the extent that the diagnosis outlined in this chapter is right, there *is* a core problem with the growing contractualisation of environmental policies.

To elaborate on this, I have made a case for the analytic usefulness of the notion of the public. I have described the public as a contingent and revisable acknowledgement of, and engagement with, otherness: a broader "us" that eschews comprehensive descriptions, a meaningful difference that impinges on action. Thus, whatever the design and implementation tactics, contractualised policies are basically at odds with publicness because they promote reciprocal immunisation and insensitivity. The same happens with the alleged capacity of the market to translate non-monetary values into monetary ones and with any claim about comprehensive commitments and accounts, as implied in much technocratic command-and-control regulation.

In conclusion, it seems to me that the public space expands when two conditions are met: no fixed terms of reciprocation (asymmetry) and no fixed ties (openness). The resulting type of relationship can be named *covenant* – provided that by covenant one means precisely a forward-oriented pledge open to further actors or acts and devoid of any specific obligation to reciprocate[14] – and distinguished from three others: *contract* (closure and symmetry); *reciprocity* (openness and symmetry); and *domination* (closure and asymmetry). Thus defined, the covenant logic, which the idea of the public draws on, bears a resemblance to Bataille's notion of *dépense*. It can be found at work in some forms of gift: the non-reciprocal, open, "first' gift" (Simmel, 1908) with which one "gives something for nothing" (Gouldner, 1973) – a blood donation; the mother who feeds her baby; the attribution of an existence value to environmental goods independently of present or future uses; and the intentional payment of avoidable costs that lies at the core of social movements and of some successful governance experiences (Frey, 1997). These forms of gift help strengthen or renew the social tie, going beyond the chains of reciprocal obligations typical of traditional societies and the closed, symmetrical relationships of modern contracts. They also avoid the self-referential asymmetry, the non-returnable unilateral gift that one finds in corporate philanthropy and humanitarian aid (Latouche, 1992), in the impersonal performance of expert systems and in other gestures of sovereign benevolence.

It is probably here, in this reservoir of sociality, that an alternative to the deadlocks of contract and the appeal to comprehensive, metaphysical communities should be sought. This is obviously only a rough suggestion of a line of research. Environmental challenges are likely to require deep social changes, yet the conditions and opportunities for the latter are difficult to identify. To this end, I believe, the notion of the public may prove valuable.

References

Baland J.M., Bardhan P., Bowles S. (eds.), *Inequality, Cooperation, and Environmental Sustainability*, Princeton (NJ), Princeton University Press, 2007.

---

[14] I follow here in part the studies of Elazar on the covenant tradition in politics (e.g. 1997).

Bendell J., Kearins K., "The Political Bottom Line: the Emerging Dimension to Corporate Responsibility for Sustainable Development", *Business Strategy and the Environment*, 14, 2005, pp. 372-383.

Boltanski L., Thévenot L., *De la justification*, Paris, Gallimard, 1991.

Boltanski L., Thévenot L., "The Sociology of Critical Capacity", *European Journal of Social Theory*, 2, 3, 1999, pp. 359-377.

Börkey P., Lévêque F., "Voluntary Approaches for Environmental Protection in the European Union – A survey", *European Environment*, 10, 1, 2000, pp. 35-54.

Boström M., "Regulatory Credibility and Authority through Inclusiveness: Standardization Organizations in Cases of Eco-labelling", *Organization*, 13, 3, 2006, pp. 345-367.

Cashore B., Auld G., Newson D., "Legitimizing Political Consumerism: the Case of Forest Certification in North America and Europe", in Micheletti M., Follesdal A., Stolle D. (eds.), *Politics, Products, and Markets*, New Brunswick, Transaction Publishers, 2004, pp. 181-199.

Dewey J. *The Public and its Problems*, New York, Holt, 1927.

Elazar D., *Covenant and Constitutionalism. The Covenant Tradition in Politics*, 3, New Brunswick (NJ), Transaction Publishers, 1997.

Esposito R., *Immunitas. Protezione e negazione della vita*, Turin, Einaudi, 2002.

Esposito R., *Bios. Biopolitics and Philosophy*, Minneapolis, University of Minnesota Press, 2008.

European Commission, *Study on Voluntary Agreements Concluded between Industry and Public Authorities in the Field of the Environment*, final report, Brussels, European Commission, 1997.

European Environment Agency, *Environmental Agreements: Environmental Effectiveness*, Copenhagen, European Environment Agency, 1997.

Falkner R., "Private Environmental Governance and International Relations: Exploring the Links", *Global Environmental Politics*, 3, 2, 2003, pp. 72-87.

Fiorino D., "Rethinking Environmental Regulation: Perspectives on Law and Governance", *Harvard Environmental Law Review*, 23, 2, 1999, pp. 441-469.

Freudenburg W., Gramling R., Davidson D., "Scientific Certainty Argumentation Methods (SCAMs): Science and the Politics of Doubt", *Sociological Inquiry*, 78, 1, 2008, pp. 2-38.

Frey B., *Not Just for the Money*, Cheltenham, Elgar, 1997.

Frynas J., "The False Developmental Promise of Corporate Social Responsibility: Evidence from Multinational Oil Companies", *International Affairs*, 81, 3, 2005, pp. 581-598.

Funtowicz S., Ravetz J. (eds.), "Post-Normal Science", *Futures*, 31, 7, special issue, 1999.

Geuss R., *Public Goods, Private Goods*, Princeton (NJ), Princeton University Press, 2003.

Godbout J., *L'esprit du don*, Paris, La Découverte, 1992.

Gouldner A., *The Importance of Something for Nothing. For Sociology. Renewal and Critique in Sociology Today*, New York, Basic Books, 1973.

Gulbrandsen L., "Organizing Accountability in Transnational Standards Organizations: the Forest Stewardship Council as a Good Governance Model", in Boström M., Garsten C. (eds.), *Organizing Transnational Accountability*, Cheltenham, Elgar, 2008, pp. 61-79.

Gunningham N., Grabosky P., *Smart Regulation. Designing Environmental Policy*, Oxford, Oxford University Press, 1998.

Gurevitch Z., "The Other Side of Dialogue: on Making the Other Strange and the Experience of Otherness", *American Journal of Sociology*, 93, 5, 1988, pp. 1179-1199.

Harrison K., "Talking with the Donkey: Cooperative Approaches to Environmental Protection", *Journal of Industrial Ecology*, 2, 3, 1999, pp. 51-72.

Heller M., Eisenberg R., "Can Patents Deter Innovation? The Anticommons in Biomedical Research", *Science*, 280, 1998, pp. 698-701.

Jacques P., Dunlap R., Freeman M., "The Organisation of Denial: Conservative Think Tanks and Environmental Scepticism", *Environmental Politics*, 17, 3, 2008, pp. 349-385.

Kerwer D., "Watchdogs Beyond Control? The Accountability of Accounting Standards Organizations", in Boström M., Garsten C. (eds.), *Organizing Transnational Accountability*, Cheltenham, Elgar, 2008, pp. 98-113.

Kimerling J., "Corporate Ethics in the Era of Globalization: the Promise and Peril of International Environmental Standards", *Journal of Agricultural and Environmental Ethics*, 14, 4, 2001, pp. 425-455.

Kollman K., Prakash A., "Green by Choice? Cross-national Variations in Firms' Responses to EMS-based Environmental Regimes", *World Politics*, 53, 3, 2001, pp. 399-430.

Konan S., Cohen M., "Information as Regulation: the Effect of Community Right to Know Laws on Toxic Emissions", *Journal of Environmental Economics and Management*, 32, 1997, pp. 109-124.

Ku A., "Revisiting the Notion of 'Public' in Habermas's Theory – Toward a Theory of Politics of Public Credibility", *Sociological Theory*, 18, 2, 2000, pp. 216-240.

Latouche S., *L'occidentalisation du monde*, Paris, La Découverte, 1992.

Laufer W., "Social Accountability and Corporate Greenwashing", *Journal of Business Ethics*, 43, 3, 2003, pp. 253-261.

Luhmann N., *Ökologische Kommunikation*, Opladen, Westdeutscher, 1986.

Luhmann N., "Observing Re-entries", *Graduate Faculty Philosophy Journal*, 16, 1993, pp. 485-498.

Luhmann N., "Limits of Steering", *Theory, Culture and Society*, 14, 1, 1997, pp. 41-57.

Lyotard J. F., *Le différend*, Paris, Minuit, 1983.

Micheletti M., *Political Virtue and Shopping*, New York, Palgrave MacMillan, 2003.

Micheletti M., Follesdal A., Stolle D., "Introduction", in Micheletti M., Follesdal A., Stolle D. (eds.), *Politics, Products, and Markets*, New Brunswick, Transaction Publishers, 2004, pp. ix-xxvi.

National Bioethics Advisory Committee, *Research Involving Human Biological Materials: Ethical Issues and Policy Guidance*, Rockville, MA, National

Bioethics Advisory Committee, 1999, Available at: http://www.georgetown.edu/research/nrcbl/nbac/hbm.pdf.

Newell P., "Citizenship, Accountability and Community: the Limits of the CSR Agenda", *International Affairs*, 81, 3, 2005, pp. 541-557.

Ostrom E., Gardner R., Walker J., *Rules, Games, & Common-Pool Resources*, Ann Arbor, University of Michigan Press, 1994.

OTA, *New Developments in Biotechnology: Ownership of Human Tissues and Cells – Special Report*, Office of Technology Assessment, OTA-BA-337, Washington (DC), US Government Printing Office, 1987.

Pellizzoni L., "The Antinomy of Accountability", in Boström M., Garsten C. (eds.), *Organizing Transnational Accountability*, Cheltenham, Elgar, 2008, pp. 210-230.

Pellizzoni L., "Revolution or Passing Fashion? Reassessing the Precautionary Principle", *International Journal of Risk Assessment and Management*, 12, 1, 2009, pp. 14-34.

Power M., *The Audit Society. Rituals of Verification*, Oxford, Oxford University Press, 1997.

Prakash A., "Responsible Care: an Assessment", *Business and Society*, 39, 2000, pp. 183-209.

Prakash A., "Green Clubs and Voluntary Governance: ISO 14001 and Firms' Regulatory Compliance", *American Journal of Political Science*, 49, 2, 2005, pp. 235-248.

Prakash A., Kollman K., "Policy Modes, Firms and the Natural Environment", *Business Strategy and the Environment*, 13, 2004, pp. 107-128.

Rai A., Eisenberg R., "Bayh-Dole Reform and the Progress of Biomedicine", *Law and Contemporary Problems*, 66, 1-2, 2003, pp. 289-314.

Sarewitz D., "How Science Makes Environmental Controversies Worse", *Environmental Science & Policy*, 7, 2004, pp. 385-403.

Schmitt C., *Politische Theologie. Vier Kapitel zur Lehre von der Souveränität*, München-Leipzig, Duncker & Humblot, 1922.

Sen A., *On Ethics and Economics*, Oxford, Blackwell, 1987.

Simmel G., *Soziologie. Untersuchungen über die Formen der Vergesellschaftung*, Leipzig, Duncker & Humblot, 1908.

Steinzor R., "Reinventing Environmental Regulation: the Dangerous Journey from Command to Self-control", *Harvard Environmental Law Review*, 22, 1, 1998, pp. 103-202.

Strange S., *The Retreat of the State: the Diffusion of Power in the World Economy*, Cambridge, Cambridge University Press, 1996.

Taylor C., *Modern Social Imaginaries*, Durham, Duke University Press, 2004.

Teubner G., "De Collisione Discorsuum: Communicative Rationalities in Law, Morality and Politics", *Cardozo Law Review*, 17, 4-5, 1996, pp. 901-918.

Teubner G., "The King's Many Bodies: the Self-deconstruction of Law's Hierarchy", *Law and Society Review*, 31, 4, 1997, pp. 763-787.

Teubner G., "Breaking Frames. Economic Globalization and the Emergence of *Lex Mercatoria*", *European Journal of Social Theory*, 5, 2, 2002, pp. 199-217.

Tovey H., "'Collective' and 'Personal' Environmentalism. Implications for Democracy of the Greening of Citizenship", in this book, *cf. supra, Environmental Democracy Facing Uncertainty*, Brussels, PIE Peter Lang, 2012.

Van den Burg S., Mol A., "Making it All Publicly Available: Four Challenges to Environmental Disclosure", in Boström M., Garsten C. (eds.), *Organizing Transnational Accountability*, Cheltenham, Elgar, 2008, pp. 177-193.

Vogel D., *The Market for Virtue*, Washington, Brookings Institution Press, 2006.

Weintraub J., "The Theory and Politics of the Public/Private Distinction", in Weintraub J., Kumar K. (eds.), *Public and Private in Thought and Practice*, Chicago, University of Chicago Press, 1997, pp. 1-42.

Wiener J., Rogers M., "Comparing Precaution in the United States and Europe", *Journal of Risk Research*, 5, 4, 2002, pp. 317-49.

Partial Knowledge in the Knowledge Society

A Case Study of an Extreme Weather Disaster and the Mitigation of Climate Change[1]

Raymond MURPHY

Emeritus Professor, University of Ottawa, Canada

One current of thought about environmental problems and the risk of disasters makes predictions of ecocide if business as usual and present trends continue. It does this by marshalling evidence of limits to growth, depletion of resources, pollution, population growth, etc. (Meadows *et al.*, 1972; Devall and Sessions, 1985; World Commission on Environment and Development, 1987; Foreman, 1991; Schneider, 1997; Ehrlich and Ehrlich, 1998; Odum and Odum, 2001; Meadows *et al.*, 2004). Proponents of an opposing paradigm predict, on the contrary, a cornucopian future with environmental problems solved. It arrives at this prediction by extrapolating present trends and constructing time series to show that the market, technology and human reason (the ultimate resource) are producing prosperity for all countries and will continue to do so if allowed to function (Simon, 1981, 1995a, 1995b, 1996; Simon and Kahn, 1985; Ray and Guzzo, 1990, 1993; Maduro and Schauerhammer, 1992; Ferry, 1992a, 1992b; Bailey, 1993; Easterbrook, 1995; Beckerman, 1995; Huber, 1999; Lomborg, 2001, 2008). Interpreting these opposing theories as plural, contested discourses valid within their respective cultural paradigms, and just leaving it at that, amounts to a rather superficial assessment. These discourses result in diametrically opposed material practices, for example carbon taxes and cap-and-trade mechanisms to reduce fossil-fuel emissions versus tax incentives to drill for more oil in nature reserves and extract oil from tar sands, which have very different ecological and socioeconomic consequences. The dis-

1 I wish to express my gratitude to the Social Sciences and Humanities Research Council of Canada for a grant that supported this research. See Murphy (2009) for the methodology and further results and analysis. Italics and indentation are used to present quotations from the leaders who were interviewed.

courses are fallible and will eventually be proven valid or erroneous by the interaction of the practices they spawn with the broader dynamics of nature.

The issue is analogous to what occurred in the World Trade Center after the two jets crashed into its twin towers. Some workers inside advocated walking up the stairwell and await rescue from the roof whereas others claimed it would be safer to walk down the 100 floors despite dense smoke. A decision had to be made in the absence of definitive knowledge, but which one? We know in hindsight that the effect of the burning jet fuel on the skyscraper support structure determined the answer to that question: the choice of going up led to death, the decision to go down resulted in safety. The crucial issue is whether socially constructed discourse and associated practices are in harmony with the dynamics of nature or out of step with them. Harmony can be the outcome of reflective decisions or it could be fortuitous. Serendipitous harmony cannot, however, be relied upon, especially in modern societies that manipulate nature's dynamics so extensively. But can presumed reflective knowledge of nature's disturbances be relied upon for safety? I have proposed a third approach that draws attention to the partial character of knowledge, to discontinuities and breaking points in trends and extrapolations (Murphy, 1994) and to unperceived or opaquely perceived risk (Murphy, 1999). The present study will examine how key decision-makers deal with uncertainty in modern society's interaction with nature's forces where knowledge is only partial.

Partial knowledge of nature's dynamics

It is necessary to probe deeper than just the risks that are talked about (Bhaskar, 1989; Dunlap and Catton, 1979, Benton, 2001a, 2001b; Dickens, 2001, 2003, 2004; Sayer, 1997, 2001; Garcia, 2004). Material problems can exist based on the autonomous dynamics of nature, before risk discourse begins and before threats are acknowledged or even perceived. These constitute latent social problems. As Adam (1998) and Dickens (2001: 93-4) have shown with respect to BSE-CJD, there is often a time lag between the beginning of nature's dynamics producing material destructiveness and social constructed conceptions of risk. Visibility of physical effects is a variable that accumulates over time and influences assumptions of safety or risk, which are social constructions forged in a social *and* a biophysical context. Current conceptions of nature only grasp some of its levels (Soper, 1995: 158), yet proper concern about nature presupposes true beliefs about it (O=Neill, 1994: 27). Nature is layered (Benton and Craib, 2001: 125): causes often lie beneath surface appearances and some knowledge claims are more valid than others (Rosa, 1998). This layering of nature, as well as the time lag between the start of nature's processes and of visible effects (visible even with science) result in unperceived risk of things that will strike.

I have argued previously (Murphy, 2002b) that as population grows and technology is developed, nature's autonomous dynamics are no longer out there in pristine wilderness but are instead internalised into society. Technology consists of "recombinant nature" or nature splicing to accomplish particular goals (Murphy, 2002b), but, as Turner (1978) has shown, those forces of nature thought mastered at times slip their leash with disastrous results. "Primal" forces of nature (Murphy, 2002b, 2004) – such as hurricanes, volcanoes, earthquakes, etc. – continue, and societies have been constructed in their path. Many of these primal disturbances are "naturogenic", but now some are "anthropogenic" (inadvertently unleashed by human actions), such as unusual flooding because of the construction of levees and dikes, more intense forest fires because of fire suppression, global warming resulting in sea-level rise and extreme weather events, etc. (Murphy, 2006a). Socially constructed movements of human agents are in a dance with the movements of nature's actants (Murphy, 2004). Whether the movements are synchronised or out of step determines whether the results are benign or disastrous. Scripts are written to prepare the dance for anticipated extreme disturbances of nature, but nature's surprises dictate that improvisation is necessary too. Since human agents are embodied beings embedded in a natural world, the movements of nature's dynamics constitute "prompts" that influence social practices and discourse. The most important types can be classified as follows (Murphy, 2004): i) "everyday prompts" from typical processes of nature; ii) "extreme prompts" from unusually severe disturbances of nature; and iii) "scientific prompts" from discoveries based on imaginative social constructions of scientists that make nature's recondite dynamics visible (see Latour, 2000). Thus discourse cannot be explained solely on the basis of prior discourse. Material contingencies also have important effects concerning which discourse will be socially constructed. Learning from material consequences is nevertheless problematic because it can require economic cost, cultural change or change in life style. Risk can be ignored or denied. Prompts can incite change, but they can also be missed or dismissed. Disaster studies (Turner, 1978; Perrow, 1984; Vaughan, 1996) have documented the social constructions that have led prompts from nature's dynamics to be ignored with disastrous consequences. Disaster itself is a prompt to new practices and beliefs, but whether the prompt leads to change depends on sociocultural *and* biophysical factors, including the severity of the disaster. Biophysical prompts that influence discourse, as well as the differing material consequences of opposing discourses, must be taken into account in the analysis rather than ignored (bracketed).

The assumption that greater knowledge of nature's dynamics results in increased control over nature is a dubious oversimplification. Some kinds of knowledge have not produced control over nature, e.g., astronomy, meteorology. The premise that an increase in knowledge reduces ignorance assumes that nature is a finite, closed system with no emerging properties. But nature is almost certainly an infinite, open system with emerging processes (Murphy, 1994). Hence we must accept the counterintuitive idea that an increase in knowledge about nature does not reduce ignorance of nature. Knowledge of nature is not mutually exclusive with ignorance of nature. Increases in scientific knowledge typically reveal new puzzles and show that the unknown is vaster than expected.

Pseudo-knowledge of nature's dynamics has also been socially constructed (we think we know at one point in time but are proved wrong later). And there remains a high level of ignorance in the classical sense (we find out by suffering the consequences that there are processes of nature we did not even imagine). Disasters in particular make visible pseudo-knowledge and ignorance. Turner (1978) conceptualises prior false expectations as a "failure of foresight" in the "incubation of disaster". Perrow (1984) and Vaughan (1996) documented that erroneous assumptions and mistakes are typical features of disasters. Such failure has even led to the collapse of societies (Diamond, 2005). The character of the correspondence between social practices and nature's forces is known retrospectively, but it is often unknown or imperfectly known in advance. Evidence is only suggestive rather than definitive.

Present trends have breaking points, so extrapolations of time series are poor substitutes for an understanding of the processes involved. Understanding nature's processes is incomplete. Partial knowledge enables the development of technology that shrinks space, brings comforts and diminishes manual labour, but it also brings new dangers from poorly understood dynamics of recombinant nature and anthropogenic primal nature (Murphy, 2004). Thus Callon *et al.* (2001) conclude that society is as uncertain as the non-human entities with which it shares its destiny.

Gross (2003: 42) argues that unawareness can paradoxically be the result of knowledge: "New knowledge and new inventions always create new non-knowledge, that is, ever new gaps in knowledge which lead to further unintended consequences." He contends that this has to be taken into account in social theory not "as mistakes and signs of incompleteness, but as probable and perhaps even unavoidable. The relation between knowledge and its side effects via non-knowledge must be considered as constitutive for one another" (Gross, 2003: 53). He distinguishes "non-knowledge" from ignorance as follows. "The term non-knowledge is a literal translation of the German *Nichtwissen*, which denotes that there can be knowledge (*Wissen*) about what is not known" (Gross, 2003: 54 fn. 2). We now know that we do not know things about nature, which we previously never imagined.

Global climate change provides an illustration. There is a scientific consensus that human activities are causing global warming, but even scientists cannot predict the specific outcomes, hence they use the inclusive and imprecise concept of climate change. Global warming could even cause dangerous cooling in some places and periods. Although it threatens to bring harm, including the risk of irreversible devastation, it could bring some benefits. Scientists have knowledge of part of the causal chain of global warming, but they also know they do not know much about the linkages and effects that will be unleashed in the future.

I would argue that non-knowledge in Gross' sense is a particular type of ignorance and incompleteness rather than being different from them in kind. The probable and the unavoidable are not polar opposites of mistakes and incompleteness. Mistakes can in some conditions be probable and the incompleteness of knowledge can in some situations be unavoidable, as Turner (1978) showed in his study of the incubation of disasters, as Perrow (1984) demonstrated in his analysis of "normal accidents" and as Vaughan (1996) documented in her analysis of the Challenger calamity. Knowledge produces unintended and unwanted consequences, and we now know this will occur. This confirms that knowledge is partial.

Beck (1998: 90, 95) argues that unawareness, not knowledge, is the medium of reflexive modernisation and a central issue is whether the inability to know becomes the justification for speeding ahead or the reason for slowing down. Is reflexive modernity taking into account this unawareness in its actions? Giddens (1994: 220) hypothesises that "the bigger the potential disaster, the more likely governing authorities and technical specialists are to say that it 'cannot occur'". This is not very reassuring when taken in conjunction with Beck's (1995: 25) assertion that "the more emphatically its possibility [of disaster] is denied, the sooner, more destructively, and more shockingly it occurs", that is, risks can be diminished only if they are accurately perceived and acknowledged. Giddens (1994) and Beck (1992) thus tend to find fault with risk-management institutions, whereas Luhmann (1993) sees them as entangled in the contradictions between systems (e.g., some systems employ the language of probability whereas others use the language of consequences). Risks are externalised from one system to another to be managed. For Luhmann (1989), effective risk management involves successful communication between systems. For Beck (1995) and others

(Stern and Fineberg, 1996; Rosa and Clark, 1999), it requires democratisation and greater public involvement.

Modern societies may be moving much more blindly in their interaction with nature's processes than they care to admit, especially if they unleash new dynamics of nature such as global climate change, which the available evidence suggests could provoke more intense and frequent extreme weather events (IPCC, 2001; Milton and Bourque, 1999: 82; Webster *et al.*, 2005). Harbingers of danger already exist for creeping global climate change, such as the melting of glaciers and of the Arctic ice cap. The question is whether prompts will be taken seriously and acted upon. Awareness that non-knowledge, pseudo-knowledge and ignorance are abundant is not an excuse. To be taken by surprise cannot serve any more as a sufficient excuse for the prolonged and protracted negligence of long-term adverse trends and equally long-term passivity (Rosenthal, 1998: 152). A crucial societal issue is: do leaders know the limitations of the knowledge base upon which they make decisions, and what will they decide if they know they do not know?

Partial knowledge and disastrous material experience

In January 1998 intense, persistent freezing rain created an ice loading on all structures that crushed a large part of the electrical grid in Quebec and the State of Maine, as well as parts of the grids of New York State and Ontario. It resulted in the most expensive disaster in Canada's history, affecting the most people, and it was the biggest natural disaster in Maine's recorded history. This long ice storm affecting an immense territory was caused paradoxically by a distant source of warming: the El Nino phenomenon in the Pacific Ocean produced warm, moist air that travelled across the North American continent and collided with the stagnant cold air mass that is normal for Northeastern North America in January. Although not caused by global warming, it may be a harbinger of extreme weather hazards when such warming occurs. I interviewed key political and emergency management leaders who had to manage the response to this extreme weather event. Did decision-makers know about the oncoming disturbance of nature in advance, or did they not perceive the risk of such a dangerous occurrence? The answer given by Florent Gagné, Deputy Minister of Public Security of Quebec, is typical of leaders I interviewed:

> Freezing rain like that was never foreseen and envisaged, even in our worst scenarios. It was certainly thought almost impossible. Preparations are always made for events that occurred in the past and that manifest a certain statistical recurrence. They are not made for the worst case, because the historical situation demonstrates that the worst case does not occur statisti-

> cally. I am convinced that in the minds of the Montreal administrators of the infrastructure, such a situation could not happen. But it happened.

Unawareness and unperceived risk resulted from unforeseen dynamics of nature. Knowledge was incomplete and predictive tools were inadequate (see also Murphy, 1999). These decision-makers who suffered disaster learned the fallibility of extrapolating safety from present trends or time series and of discourse predicting safety. They discovered that modern means of reflective risk assessment had failed to foresee the severity of nature's disturbance. They now have a better appreciation of how little is known about disturbances of nature.

What do decision-makers decide when they know they do not know? Mr. Gagné replied that:

> Science is quite limited for the prediction of events of that type. There is a tendency after experiencing a disaster to prepare for the disaster that was just experienced and not for the next one.

The material experience of disaster was a powerful contingency affecting perceptions and practices, but these leaders realised that the next hazard to strike remains unknown.

The co-construction of a hybrid forum

With all the uncertainties concerning extreme weather events, how was learning from this one accomplished? After the storm ended and power was restored, the worst-hit region – Quebec – created a commission of inquiry into the disaster. Mr. Gagné argued that:

> The choice of President says it all. If a judge is chosen, then a guilty party is sought. Condemning someone is the goal. Whereas here, what was needed wasn't to determine guilt, but rather to find solutions to help us in the future.

The President of the order of engineers was named as President of the commission. Its members included a former dean of a university faculty of science and engineering, a high ranking public servant, a lieutenant-general in the military, a mayor of a city, the vice-president of the human rights commission of Quebec, as well as a sociology professor specialising in disasters and risk management. The commission was made independent of government and of the power utility. It had a substantial budget to order scientific studies of the disaster and to hold public meetings to allow input from concerned groups. Collaboration between technical experts, political representatives and concerned laypeople was fostered to determine what is known and what should be investigated. This combination of the best available scientific research and highly visible public democracy was the self-reflexive means the state used to attempt to demonstrate that the government and the public power utility were being held accountable to the people, that lessons would be learned, and that risk would be minimised in the future. The construction of this "hybrid forum" (Callon *et al.*, 2001), discussing the boundary between the technical and the social, was effective in clarifying the issues and in reassuring the population in the context of uncertainties engendered by primal nature's construction of freezing rain in an electrically dependent and vulnerable society (Commission, 1999: xiii-xv).

Callon *et al.* (2001) contend that the question of the conditions under which such collaboration between specialists and laypeople comes to the fore is crucial in a technical democracy. The prompt of a disaster is part of the answer. The social construction of this hybrid forum can only be understood by taking into account the material contingency of the extreme prompt provided by nature's construction of this weather event. The hybrid forum was a co-construction of the political culture of Quebec and of nature's crushing of the essential infrastructures of this modern society. The temporary replacement of normal weather by an extreme weather event incited a social event in which routine democratic practices were provisionally replaced by enhanced democratic practices. The creation of a hybrid forum is not unique to Quebec. I would hypothesise that its social construction has been typically prompted by the experience of disaster.

Assessing risk and preparing for disaster

The commission entitled its five-volume report *Confronting the Unforeseeable* (Commission, 1999), thereby confirming the salience of partial knowledge and non-knowledge: modern society and its technical experts know they do not know. In the fog of risk assessment, the organisational structure orchestrated to deal with disasters is problematic. Should it be one all-hazards organisation or several focused, hazard-specific organisations? Maine's Governor Angus King replied that:

> The problem with preparing for disasters is you never know what the disaster is going to be. I definitely think you could do it with the same organisation. You'd have to have different capacities and tools to cope with what the risks are.

In Quebec, too, the solution is to make available specific organisational tool kits necessary for very diverse hazards within one all-hazards, cost-effective organisation. The Director of the Montreal Centre for Civil Security, Jean-Bernard Guindon, stated:

> Our decision here is to have a generalist organisation to deal with almost anything. However, that organisation must have specific intervention plans and particular approaches concerning certain types of calamities. Seven sce-

narios of calamities have been identified for which approaches have been prepared in decision making for all events that could occur during those scenarios.

This requires accurate identification of risk, in particular, that all the hazards fall into one of the seven foreseen scenarios. That is a cost-effective working hypothesis, but previous disturbances of nature invalidated risk assessments and partial knowledge that existed at the time, in particular prior to the 1998 extreme weather event. A deeper lesson of this disaster is that an eighth unforeseen scenario could also occur, nature can surpass expectations of hazards discounted by economics and life-style habitus, and hence to expect the unexpected.

A disastrous extreme prompt has an immediate impact on perceptions, discourse and practices, but the duration of the impact varies according to social location. In the wake of this disaster, Florent Gagné concluded that the population's concern with disaster preparation wanes as time passes after the extreme prompt and that politicians change so much that those who dealt with the disaster are no longer in office. Hence learning from a disaster is almost entirely institutional, located in the professional class of disaster experts, disaster organisations and legislative changes:

> For professionals of public security, the culture [of security] does not wither away. On the contrary it is being reinforced and organised more and more. On the other hand in terms of politicians and the population, I don't have any indication at all of a culture of civil security that is stronger and more developed now than it was before the ice storm. It is strictly at the level of disaster professionals where there was definite and considerable progress.

Modern society has become highly dependent on expert organisation for disaster preparation.

These decision-makers argue that it is too expensive to prepare 100% for disasters and it probably would not be possible anyway because we do not know what nature will throw at us. Mr. Gagné's articulation of this is typical of all the other leaders:

> It would be unreasonable to prepare for the maximum worst case. It is unfortunately necessary to take a certain risk and try to minimise it by an efficient organisation, a good deployment when something happens, and then count on a well-organised society, a rich society in which we live to have the means.

This implies acceptable risk, which means accepting disaster in the hope that its effects can be minimised and its frequency will be low.

Everyday and disaster decision-making

Although it may seem surprising, leaders find it easier to make decisions during a disaster because that certain knowledge of harm leads the population to be more accepting of painful remedies and to rally round the decision maker. Hubert Thibault, Chief of Staff for Quebec Premier Bouchard, stated that:

> The most difficult decision was not in the immediate management of the crisis. A crisis situation evokes crisis decisions. People understand that. Following the ice storm crisis, the level of satisfaction in the government of Mister Bouchard rose to 90%.

The sense of urgency incited by a disaster does not prevail under normal conditions of nature. The chronic task of prevention of disasters and environmental problems is much more difficult in the context of partial knowledge and uncertainty when everyday prompts from nature lead to a feeling of well-being. He argued that:

> The most difficult decisions occurred when the immediate crisis began to dissipate, when decisions had to be made to ensure it would not occur again. Decisions were not difficult to make in terms of reasoning because we were convinced they were things that had to be done, but they were more difficult to sell.

Prevention and preparation involve decisions that irritate because they conflict with other priorities. He pointed out that "security does not just consist of buying fire trucks. It involves many things that taken individually cause irritation, but are necessary".

The major irritant was the construction of an additional power transmission line, which faced opposition because no one wanted it in their neighbourhood, even though everyone would love to have it somewhere else. Moreover, risk assessments in times of normality are based on the partial knowledge of probabilities.

Maine's Governor Angus King faced similar resistance:

> The folks who are concerned about climate change have to understand there is no free lunch when it comes to energy. There's a major wind project proposed in Western Maine and it's being fought to the death by the Appalachian Mountain Club because – heaven forbid – a hiker walking through the mountains would see a windmill and spoil his wilderness experience. These are the same people who bemoan fossil fuels, but don't put windmills on my hiking trail. That absolutely drives me crazy. Every choice has consequences. If you don't want to do fossil fuels, then you have got to accept windmills and dams and maybe even nuclear power.

If technological development and the population's choices result in more energy consumption – for example, replacing cathode ray televisions with large flat screen plasma TVs that draw more power – then the energy must come from somewhere. Both conservation and supply of clean energy face opposition. Decisions to promote sustainability face resistance when lurking threats are only partially visible and knowledge is incomplete. This contrasts with decisions to restore normality while the effects of disaster are felt and constitute certain knowledge.

The shock of a sudden disaster and the slow onset of global warming

What interpretations did leaders who managed this extreme weather disaster construct about global climate change and how to deal with it? The Mayor of Montreal, Pierre Bourque, argued that the population and leaders have to experience shocks such as disasters in order to become sensitised to risk: "Otherwise people live fundamentally without stress. Why be bothered with unhelpful stresses?"

Knowledge remains theoretical until confirmed by material experience. So I asked how a phenomenon like global climate change is dealt with, given there are no shocks. It is something that creeps up slowly and risks changing our world completely and irreversibly before there is a shock. How do people learn in situations where there are no advance shocks? Mayor Bourque could only express his worries that societies may not adapt in time.

Is it necessary to suffer a disaster to become motivated to prepare for disaster? Hubert Thibault, Chief of Staff for the Premier of Quebec, responded: "There is a dimension of that which is true."

Global climate change is a slow-onset occurrence. It may constitute the incubation of disasters but present political decision-makers presume that disastrous extreme prompts won't happen while they are in office. Mr. Thibault contended that:

> When one reads scientific texts concerning climate, it is obvious that problems will not occur before the end of the mandate of today's politicians. I am saying something important. It will be fifty years away.

Political decision-makers would receive the blame for imposing sacrifices necessary to solve slow-onset environmental problems such as climate change but will be out of office before beneficial results are visible and credit awarded. Hence they are predisposed to discount future problems (much like economists discount future costs) and leave human activities underlying global warming as they are to avoid irritating voters.

In the absence of extreme prompts of disaster, scientific prompts coming from organisations such as the IPCC have to move against the strong current of everyday prompts of well-being. Maine's Governor, Angus King, who was one of only two independent Governors in the USA at the time of this disaster, and neither Democrat nor Republican, gave his analysis of the way climate change is being dealt with in the United States:

> Global warming is an ideological issue among American conservatives. They don't believe it and if you believe in global warming you're a wimpy liberal, which is nonsense to me. You try to go on the science, and the majority view of the science is clearly that there is a human contribution to this problem that's significant.

He saw these conservatives as the power behind the George W. Bush administration. Governor King did not criticise that administration for refusing to sign the Kyoto Protocol, which he concluded was problematic. But he did ...:

> [...] criticise them for not seeking another alternative and for emphasising drilling as the solution to our energy problems, which is very shortsighted. There is so much energy wasted in this country, that without a hell of a lot of discomfort to anybody, we could save a huge amount and develop alternatives. The only energy policy seems to be greater subsidies to the fossil-fuel industry to produce more, which is only a palliative, a short-term solution.

The Montreal Emergency Management Director, Jean-Bernard Guindon, expressed this failure of foresight in particularly strong ethical terms.

> It is socially irresponsible not to implement the Kyoto Protocol. When I see countries like the United States that refuse, I see that as a crime against humanity because they refuse to take the means that could protect future generations against catastrophe. In the meantime, we are stuck with living with the consequences of climatic extremes.

The treadmill of consumption and production

Some of these decision-makers connected environmental problems such as global climate change to consumerism. Montreal's Mayor, Pierre Bourque, argued that:

> We consume outrageously. The Chinese will consume and the Indians will consume. I cannot be against development either. I cannot say: we will remain rich and the others will remain poor. This involves a problem of universal conscience. Materialism is so powerful. Everyone for himself. Every country for itself. It is worrisome for the future.

Mayor Bourque was educated as a botanist and before entering politics was Director of Montreal's Botanical Gardens. He advocated sustainable development but was quite worried about the impact of consumerism on the natural infrastructure needed to sustain societies.

Consumption of fossil fuels is particularly high in North America. I pointed out to Maine's Governor King that with taxes on gas being relatively low in the United States, there is no real incentive to develop alternatives. Then I asked him whether a solution to this would be to tax gas like they do in Europe. He answered that "everybody that's ever mentioned it has disappeared".

Leaders and the population pin all their hopes on the development of technological knowledge in the unfettered market to provide timely solutions when needed, a faith based approach if ever there was one.

Conclusion

There is widespread scientific consensus that human activities are contributing to global warming. Increased flooding from a rising sea-level and more frequent and intense extreme weather events are some foreseeable consequences, which could result in disasters for places that are vulnerable for geographical or social reasons. Scientists know, however, that they do not know much about global climate change. How do political and emergency-management leaders, especially those who have experienced disaster, deal with uncertainty? This study drew from some of the conceptual work of environmental sociology and disaster sociology, then used this integrated perspective to analyse interviews with decision-makers in regions of Canada and the United States struck by an extreme weather event caused by unusual warming: an ice storm.

The investigation showed how nature's dynamics affect culture, in particular how extreme and everyday prompts from nature were material contingencies that influenced the beliefs of key decision-makers who had just experienced a disaster. Material experience is a particularly important contingency in the formation of beliefs and knowledge of risks. Extreme prompts of disaster make visible erroneous extrapolations of safety. The extreme weather event incited the launching of an enhanced democratic event in the form of a hybrid forum of a commission of inquiry to gain more knowledge, but as normal weather replaced extreme weather, so too enhanced democracy gave way to routine democracy and dependence on experts. Awareness and/or acknowledgement of risk by the population and their political leaders tend to recede as the experience of well-being replaces the experience of disaster. Disaster, nevertheless, prompts a ratcheting up of organisational means of safety that are embodied in the professional class of disaster experts and institutionalised. They have the function of learning from disaster to prepare for the future. The importance of material experience has the result that society is reasonably well prepared for yesterday's disturbance of nature but doubtfully prepared for tomorrow's because knowledge of risks is partial, as admitted by these decision-makers.

This has major implications for dealing with environmental problems, such as global climate change. If the experience of leaders and the public is one of everyday prompts of well-being, then scientific prompts of risk – for example, IPCC forecasts of global climate change – tend to be trumped by predispositions to inertial habitus in influencing social practices. Safety requires accurate identification of hazards, yet these decision-makers admit they do not know what disasters will strike in the future. They do not seek maximum protection, not only because it would be too expensive but also because they know they do not have the knowledge necessary to attain it. They know they are accepting risk, hence accepting disaster, but they do not know what it will be. Although they acknowledge they do not know when disaster will occur, they presume they know that it will occur later rather than sooner and not while they are in office. These leaders find it much more difficult to promote costly safety measures needed to reduce risk in the context of material well-being than to make decisions during a disaster. This is typical of the difficulty under nature's everyday prompts to convince the public to accept decisions that irritate but are necessary to solve slow-onset environmental problems that creep up on society. These decision-makers were well aware of the lack of political will by North America federal leaders at the time of the interviews (before Obama took power in the USA) to solve the problem of greenhouse-gas emissions, in one case depicting it as a crime against humanity, and of its connection to consumerism.

The fallibility of socially constructed discourse extrapolating safety from present trends of well-being and time series was confirmed by this disaster. Rather than only perceiving the onward march of increasing knowledge, the results of this investigation draw attention to how partial our knowledge of nature's dynamics is. The study increases non-knowledge about the dynamics of nature in Gross' sense: it adds to awareness that modern societies do not know a great deal about the interaction between their technological innovations and the broader dynamics of primal nature. The application of scientific knowledge in technological development is producing new unknowns concerning nature's emergent dynamics. These decision-makers had a heightened sense of such non-knowledge after the disaster struck their area. As societies recombine nature's dynamics in novel ways and inadvertently unleash new processes of primal nature, leaders face the predicament of knowing they do not know. By creating unfamiliar vulnerabilities to autonomous forces of nature, new technologies have forced perplexing decisions on modern societies. Leaders and the population can socially construct whatever assumptions they want – those of safety or danger – but these lead to different practices with different material consequences in their interaction with nature's processes. Two sources of uncertainty abound: one based on uncertainty about what nature will construct; the other on uncertainty about which material practices will be constructed by societies. Greater awareness of unawareness implies that business as usual and full speed ahead is particularly risky. Meeting the challenge of reducing disasters in the context of global climate change requires a healthy respect for the limits of knowledge: an awareness that pseudo-knowledge, ignorance and partial knowledge are plentiful in the knowledge society. Knowing that we do not know is an important component of reflexive modernisation and implies the necessity of precautionary advance.

References

Adam B., *Timescapes of Modernity: the Environment and Invisible Hazards*, London, Routledge, 1998.

Bailey R., *Eco-Scam: the False Prophets of Ecological Apocalypse*, New York, St. Martin's Press, 1993.

Beck U., *Risk Society*, London, Sage, 1992.

Beck U., *Ecological Politics in an Age of Risk*, Cambridge, Polity Press, 1995.

Beck U., *Democracy without Enemies*, Cambridge, Polity, 1998.

Beckerman W., *Small is Stupid: Blowing the Whistle on the Greens*, London, Duckworth, 1995.

Benton T., "Environmental Sociology", *Sosiologisk tidsskrift*, 9, 2001a, pp. 5-48.

Benton T., "Why Are Sociologists Naturephobes?", in Lopez J., Potter G. (eds.), *After Postmodernism*, London, Athlone. 2001b.

Benton T., Craib I, *Philosophy of Social Science*, Hampshire, Palgrave, 2001.

Bhaskar R., *Reclaiming Reality*, London, Verso, 1989.

Callon M., Lascoumes P., Barthe Y., *Agir dans un monde incertain. Essai sur la démocratie technique*, Paris, Seuil, 2001.

Collectif, *Pour affronter l'imprévisible*, Quebec, Publications du Québec, 1999.

Delèves A., Ferry L., "Entre la nature et l'homme, je choisis l'homme", review of Luc Ferry, *La vie*, 2465, 26 novembre 1992, pp. 15-16.

Devall B., Sessions G., *Deep Ecology: Living as if Nature Mattered*, Salt Lake City, Peregrine Smith, 1985.

Diamond Jared, *Collapse: How Societies Choose to Fail or Succeed*, New York, Viking, 2005.

Dickens P., "Linking the Social and Natural Sciences", *Sociology*, 35, 2001, pp. 93-110.

Dickens P., "Changing Our Environment, Changing Ourselves", *Interdisciplinary Science Review*, 28, 2003, pp. 95-105.

Dickens P., *Society & Nature*, Cambridge, Polity Press, 2004.

Dunlap R., Catton W., "Environmental Sociology", *Annual Review of Sociology*, 5, 1979, pp. 243-273.

Easterbrook G., *A Moment on the Earth: the Coming Age of Environmental Optimism*, New York, Viking, 1995.

Ehrlich P.R., Ehrlich A.H., *Betrayal of Science and Reason: How Anti-environmental Rhetoric Threatens Our Future*, Washington (DC), Island Press, 1998.

Ehrlich P.R., Ehrlich A.H., *One with Nineveh: Politics, Consumption and the Human Future*, Washington (DC), Island Press, 2004.

Ferry L., *Le nouvel ordre écologique*, Paris, Grasset, 1992a.

Ferry L., *La vie*, 2465, 26 novembre 1992b.

Foreman D., *Defending the Earth*, Montreal, Black Rose Books, 1991.

Garcia E., *Medio Ambiente y Sociedad: la Civilizacion Industrial y los Limites Del Planeta*, Madrid, Alianza Editorial, 2004.

Giddens A., *Beyond Left and Right*, Stanford, Stanford University, 1994.

Gross M., "Sociologists of the Unexpected: Edward A. Ross and Georg Simmel on the Unintended Consequences of Modernity", *The American Sociologist*, 34, 2003, pp. 40-58.

Huber P., *Hard Green: Saving the Environment from the Environmentalists*, New York, Basic Books, 1999.

Intergovernmental Panel on Climate Change (IPCC), *Climate change 2001*, Cambridge, Cambridge University, 2001.

Latour B., "When Things Strike Back", *British Journal of Sociology*, 51, 2000, pp. 107-123.

Lomborg B., *The Skeptical Environmentalist: Measuring the Real State of the World*, Cambridge, Cambridge University Press, 2001.

Lomborg B., *Cool It: the Skeptical Environmentalist's Guide to Global Warming*, New York, Vintage, 2008.

Luhmann N., *Ecological Communication*, Cambridge, Polity, 1989.

Luhmann N., *Risk: a Sociological Theory*, New York, Aldine de Gruyter, 1993.

Maduro R., Schauerhammer R., *The Holes in the Ozone Scare: Environmentalists vs. Modelers*, Washington (DC), Twenty-first Century Science Associates, 1992.

Meadows D. *et al.*, *The Limits to Growth*, New York, Universe Books, 1972.

Meadows D., Randers J., Meadows D., *The Limits to Growth: the 30-Year Update*, White River Junction VT, Chelsea Green, 2004.

Milton J., Bourque A., *A Climatological Account of the January 1998 Ice Storm in Quebec*, Ottawa, Environment Canada, 1999.

Murphy R., *Sociology and Nature*, Boulder, Westview, 1997.

Murphy R., *Rationality and Nature*, Boulder, Westview, 1994.

Murphy R., "Unperceived Risk", *Advances in Human Ecology*, 8, 1999, pp. 99-123.

Murphy R., "Natures Temporalities and the Manufacture of Vulnerability", *Time and Society*, 10, 2001, pp. 329-348.

Murphy R., "Extreme Weather and the Energy Metabolism of the City", *Environment and History*, 8, 2002a, pp. 43-64.

Murphy R., "The Internalization of Autonomous Nature into Society", *Sociological Review*, 50, 2002b, pp. 313-333.

Murphy R., "Disaster or Sustainability", *Canadian Review of Sociology and Anthropology*, 41, 2004, pp. 1-18.

Murphy R., "The Challenge of Disaster Reduction", in McCright A., Clark T. (eds.), *Community and Ecology: Dynamics of Place, Sustainability, and Politics*, Amsterdam, JAI/Elsevier, 2006a, pp. 93-120.

Murphy R., "Environmental Realism: from Apologetics to Substance", *Nature & Culture*, 1, 2, 2006b, pp. 49-72.

Murphy R., "Thinking across the Culture/Nature Divide: an Empirical Study of Issues for Critical Realism and Social Constructionism", in Frauley J., Pearce F. (eds.), *Critical Realism and the Social Sciences*, Toronto, University of Toronto Press, 2007, pp. 142-161.

Murphy R., *Leadership in Disaster: Learning for a Future with Global Climate Change*, Montreal, McGill-Queen's University Press, 2009.

O'Neill J., "Humanism and Nature", *Radical Philosophy*, 66, 1994, pp. 21-29.

Odum H.T., Odum E.C., *A Prosperous Way Down: Principles and Policies*, Boulder, University Press of Colorado, 2001.

Perrow C., *Normal Accidents*, New York, Basic, 1984.

Ray D., *Environmental Overkill: Whatever Happened to Common Sense*?, New York, HarperCollins, 1993.

Ray D., Guzzo L., *Trashing the Planet*, New York, HarperCollins, 1990.

Rosa E., "Metatheoretical Foundations for Post-normal Risk", *Journal of Risk Research*, 1, 1998, pp. 15-44.

Rosa E., Clark D., "Historical Routes to Technological Gridlock", *Research in Social Problems and Public Policy*, 7, 1999, pp. 21-57.

Rosenthal U., "Future Disasters, Future Definitions", in Quarantelli E.L. (ed.), *What is a Disaster?*, London, Routledge, 1998, pp. 146-159.

Sayer A., "Essentialism, Social Constructionism, and Beyond", *Sociological Review*, 45, 3, 1997, pp. 453-487.

Sayer A., "Reply to Holmwood", *Sociology*, 35, 2001, pp. 967-84.

Schneider S., *Laboratory Earth: the Planetary Gamble We Can't Afford to Lose*, New York, Basic Books, 1997.

Simon J., *The Ultimate Resource*, Princeton, Princeton University Press, 1981.

Simon J., "Introductions", in Simon J. (ed.), *The State of Humanity*, Cambridge, Blackwell, 1995a, pp. 1-28.

Simon J., "What Does the Future Hold?", in Simon J. (ed.), *The State of Humanity*, Cambridge, Blackwell, 1995b, pp. 642-660.

Simon J., *The Ultimate Resource 2*, Princeton, Princeton University Press, 1996.

Simon J., Kahn H., *The Resourceful Earth*, Oxford, Basil Blackwell, 1984.

Soper K., *What is Nature?*, Oxford, Blackwell, 1995.

Stern P., Fineberg H. (eds.), *Understanding Risk*, Washington, National Academy, 1996.

Turner B., *Man-made Disasters*, London, Wykeham, 1978.

Vaughan D., *The Challenger Launch Decision*, Chicago, University of Chicago, 1996.

Webster P.J., Holland G.J., Curry J.A., Chang H.R., "Changes in Tropical Cyclone Number, Duration, and Intensity in a Warming Environment", *Science*, 309, 5742, 2005, pp. 1844-1846.

World Commission on Environment and Development, *Our Common Future*, Oxford, Oxford University Press, 1987.

Integration of Popular Knowledge in Sustainability Research

Martina SCHÄFER* AND Tina BOECKMANN**

**Professor, **Research Associate, Technical University of Berlin, Germany*

With the 1992 UNCED-Conference in Rio de Janeiro, the concept of sustainable development, bringing together questions of social justice with environmental issues, has gained importance in international politics. Based on these political processes, a new field of research – sustainability research or social-ecological research – has developed within the last decade in several European countries. Due to the complex problems it deals with and the challenge of developing options for a relatively profound social and ecological transformation of society, it is often organised as inter- and transdisciplinary research. This type of research faces the challenge both of integrating knowledge and methods from different scientific disciplines and of integrating practitioners' with scientific knowledge. The intention is to bring together perspectives, expertise and experience from various social actors, developing solutions that can easily be put into practice. Some experience has already been derived from this type of research and it can be analysed and learned from.

The experience discussed here was generated through a social-ecological research programme funded by the German Federal Ministry of Education and Research.[1]

Inter- and transdisciplinary research

In contrast to strictly disciplinary research, *interdisciplinary research* refers to a common problem, or set of problems, situated at the intersection of several disciplines. Dealing with these problems requires clearly defined and specifically organised types/forms of cooperation and the

[1] http://www.sozial-oekologische-forschung.org

linkage of theoretical perspectives and methods among the different disciplines (Brand, 2000: 14; Tress *et al.*, 2003).

In addition to these characteristics, *transdisciplinary research* refers to societal issues met in the lifeworld, in contrast to problems derived from scientific theory. To be able to approach these problems, cooperation between academic researchers and practitioners or societal actors is necessary. Working toward solutions for these problems is no longer linked to disciplinary research interests or convention-bound obligations to use certain methods within a discipline (*ibidem*, p. 15). Based on the Swiss experience with the Swiss Priority Program Environment (SPPU), Häberli *et al.* (1998) argue that in transdisciplinary research "the task is defined by the real problem and not by the competence or the analytical instruments available". Besides producing knowledge, the goal in these projects is to intervene in societal practice with the knowledge generated, thus moving from knowledge production to knowledge intervention (Nicolini, 2001). Important characteristics of this type of research are the participative development of research questions and the broad diffusion of results in societal contexts.

The rise of this type of research is seen by some authors to be an indicator of a fundamental change in science. Gibbons *et al.* (1994) talk of a new kind of knowledge production, "mode 2", in contrast to the existing disciplinary methods of knowledge production, "mode 1". Gibbons and Novotny (2000) explain this development in the following manner:

> A transformation is occurring in the relationship of science and society. The characteristics of a new mode of knowledge production – problem definition, heterogeneity, transdisciplinarity, accountability, and multiple measures of quality control – are at the heart of this transformation. Much of the thrust of innovation today is coming from new links between traditionally segmented producers and users of knowledge. The contextualisation of research around the interests of a wider range of stakeholders fosters a more "socially robust" knowledge, that transgresses disciplinary and institutional boundaries.

Other authors state that the rise of a problem-oriented research type offers "a complement to the established forms of disciplinary science" (Grunwald, 1999). For post-normal science (Funtowicz and Ravetz, 1993; Becker, 2000: 76), "mode 2" signifies that the production, systematisation and transfer of science are no longer bound to certain discipline-defined locations. Society has been taken over by science and the boundaries of science have been loosened up.

Especially in the fields of environmental and risk-research science, the aura of objectivity is disappearing. The ongoing, institutionalised conflict concerning the risks of modern high technology shows clearly that science is socially constructed. The selectivity of science – its links to economic and political interests, as well as its anchoring in world views and powerful social networks – has become evident. Non-scientific knowledge (e.g. local experimental knowledge, risk assessment by lay persons) is gaining importance in parallel with science's loss of authority (Wynne, 1988; Funtowicz and O'Connor, 1999; Brand, 2000).

Transdisciplinarity in sustainability research

Since sustainability research deals with complex societal problems, it consequently has to use interdisciplinary and transdisciplinary forms of knowledge production. These problems are characterised by being globally interlinked, complex, synergetic, cumulative and highly dynamic, often marked by non-linear causal chains and significant time lags between causes and effects in the interplay between social and natural systems (Siebenhüner, 2004: 76).

Brand (2000: 24) describes the function of transdisciplinary sustainability research as follows:

> Transdisciplinary sustainability research aims to contribute to the solution of different problems of sustainable development. The premise is that these solutions cannot be worked out by science alone nor by highly specialised disciplines. On the contrary, sustainability problems partly result from the dynamics of scientific, technical and societal processes of differentiation. Nor, however can solutions be found without highly specialised science – what is needed is the creation of new, problem-oriented, interdisciplinary connections.

In sustainable or social-ecological research, a differentiation is made between three types of knowledge generated (Cass/ProClim, 1997; Maier Begré and Hirsch Hadorn, 2004):

– *Systemic knowledge:* knowledge about the characteristics and dynamics of processes and the interconnections between ecological, economic, social and cultural aspects.

– *Normative (ethical) knowledge*: knowledge that helps to evaluate the sustainability of societal transformation.

– *Transformational knowledge:* knowledge that helps to develop strategies for the societal transformation process.

All three types of knowledge are linked with each other and need to be analysed interdependently (Hirsch Hadorn *et al.*, 2002: 15). There are different reasons for integrating non-scientific or popular knowledge into the generation of each of these types of knowledge. Regarding the generation of systemic knowledge, exchange with practitioners allows insight into interconnections and limitations that can only be experienced by being an actor in the field concerned. Including other forms of knowledge also makes it easier to obtain an impression of the different scales (local, regional, national, global) and the multiple, interactive and cumulative nature of certain problems (Siebenhüner, 2004: 77).

The need to integrate different societal perspectives seems especially obvious with respect to the generation of normative knowledge. Science can help to identify different interests and moderate the process of discussing transformation goals, but it cannot define normative premises on its own. The definition of critical loads or threshold values, the evaluation of risks, the ways of dealing with insecurity, the choice of transformation goals and indicators – all of these steps are based on normative decisions, which imply conflicts of interests and values. Since only widely accepted transformational goals will be put into practice, the aim should be the widest possible participation of those who will be affected (Brand, 2000: 20).

Concerning transformational knowledge, exchange with practitioners is essential for the development of solutions or strategies that are context-specific and will be relevant in practice. Implementation of the required changes stands a greater chance of success if those groups affected are involved not only in the definition of transformation goals, but also in the construction of transformation strategies and measures (*ibidem*).

Being involved in transdisciplinary research confronts scientists with new demands, such as being able to: use new methods and publication strategies; vary discursive register, depending on audience; or moderate integration processes. The past decade has seen a range of experiences in all these fields, some already documented and subjected to systematic interpretation. The reflective exchange of experiences makes it easier to refer to methods that are context and target-group-related (e.g. Hirsch Hadorn *et al.*, 2002; Luley and Schramm, 2003; Schophaus *et al.*, 2004; Boeckmann and Noelting, 2005) and allow for the development of criteria for ensuring the quality of transdisciplinary research.

Project experiences within the Social-Ecological Research Programme (Germany)

The Social-Ecological Research Programme (SERP) was initiated by the German Federal Ministry of Education and Research in order to "describe and analyse social-ecological problems and transformations" (Balzer and Wächter, 2002). During the last 10 years approximately 50 projects have been funded, all focused on very different sustainability problems (agriculture and nutrition, climate change, transformation of supply systems and settlements, strategies of regional sustainability, sustainable consumption, etc.). However, the common factor is that all of them are carried out by interdisciplinary research teams and in exchange with societal actors, such as businesses, government administrators or politicians, NGOs, etc. At the end of 2004 a first impression concerning the objectives for practitioner participation, the methods applied to incorporate popular knowledge, and the experiences already recorded was obtained through questionnaires and telephone interviews (Boeckmann and Schäfer, 2004; Baranek *et al.*, 2005). Some of the results are presented in the remainder of this section.[2]

Objectives and methods of participation

The objectives for exchange with societal actors in the different projects can be differentiated into:
– the collection of empirical data;
– the collection of systemic, normative, or transformational knowledge in order to generate appropriate context-specific solutions;
– the diffusion of knowledge and the implementation of transformation strategies in practice (Baranek *et al.*, 2005: 27).

While the first objective is not specific to transdisciplinary research, the researchers aim to develop better solutions through exchange with practitioners (objective 2) or seek to ensure that the transformational knowledge produced is recognised as valid by the relevant actors and integrated into societal processes (intervention) (objective 3). Additionally, the projects aim to link the participants, as well as promote dialogue, learning processes, a sense of understanding regarding sustainable development and a willingness to become active in this field.

The actors involved in participation processes have typically been consumers (users of products), entrepreneurs (farmers, producers and retailers), stakeholders (NGOs and other interest groups), political decision-makers, administrators, multipliers and consultants.

The methods used vary among interviews, surveys, focus groups and workshops, the latter at times being aimed at constructing participative scenarios or models. In one project, an advisory board with practitioners was set up. Sometimes surveys or conferences have been planned collaboratively. Experiments with innovative methods, such as joint excursions, "story telling", "fantasy trips" or the setting up of a photo studio, have featured in some of the projects (Baranek *et al.*, 2005).

Relation between objectives and intensity of participation

From an analysis of intended goals and the intensity of participation in the projects (e.g. starting point, methods applied and actors chosen), a

2 The results presented here are based on the responses from 16 projects.

clear relation emerges. In those projects that are mainly aimed at generating systemic knowledge, participation is less intense than in projects aimed at generating transformational knowledge or even at analysing a specific intervention in practical processes (Boeckmann and Schäfer, 2004). When systemic knowledge is being generated, the participating actors mainly have the task of delivering information about the structure, dynamics and decision patterns in the field of interest (e.g. agriculture, climate policy, etc.). Establishing relations of trust among different actors in the field can be useful in providing insights and additional information. It can, however, be a process that takes a great deal of time; therefore, a less intense participation process does not always mean less input or effort on the part of the research team.

Participation becomes more intense when the generation of transformational knowledge is sought and when the project seeks to ensure acceptance and transfer in the field of interest. For example, in one of the projects, the researchers are trying to ensure the transfer of knowledge concerning mobility-related lifestyles and options for behavioural change by participating for six months in the work of a number of mobility services.

The highest intensity of participation and the widest variety of methods applied is found in projects aimed at intervention into societal processes using the generated transformational knowledge. To understand participation in these projects, the actors need to be seen as far more than a partner for interviews or a deliverer of information and data. Rather, the projects are aiming for a relationship among "equal partners", right from the very beginning, when formulating the research questions. At that point, the nature of the collaboration and the roles of the partners involved can be defined. For projects seeking to intervene in societal processes, a different set of methods and close contact with the actors in the field is necessary, since the successful transfer of knowledge is one of the main criteria upon which the success of the project will be evaluated. When aiming for a common goal and a joint definition of the problem, an already active collaboration with some of the central societal actors has been shown to pay off during the application phase. The results of the 2004 survey show that the research objectives and the importance given to participation determine the methods and the points of integration of popular knowledge (*ibidem*). These relations are illustrated in the following diagram:

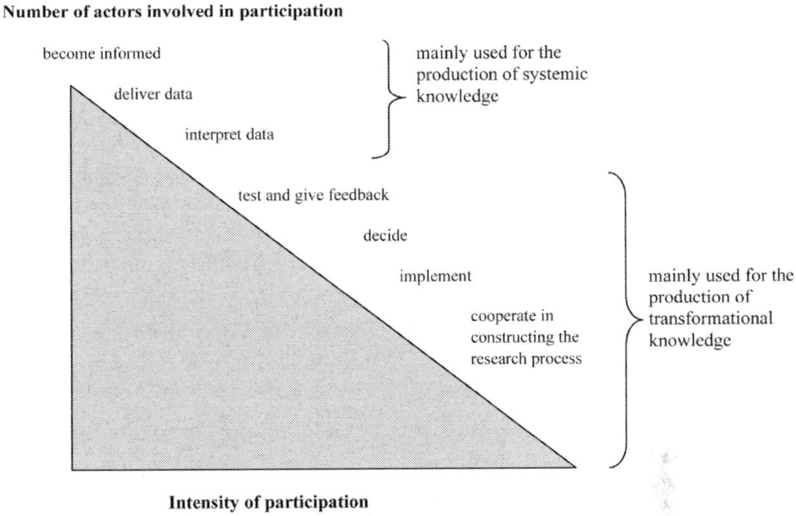

Figure 1 – Relation between the number of actors involved in participation, their function and the intensity of participation

Objectives and methods of transdisciplinary research in the project "Regional Wealth Reconsidered"

"Regional Wealth Reconsidered. The Contribution of the Organic Agriculture and Food Industry toward Quality of Life" (RWR) was a SERP project dealing with questions in the fields of agriculture and nutrition in the years 2002 to 2007.[3] It was carried out by an interdisciplinary research team and sought to integrate popular knowledge from different actors at various stages of the project. In this section (4.1), the project's objectives and methods of participation are presented and the experience derived from it is discussed. Section 4.2 discusses the effects that can be obtained from this type of research.

Participation methods used

Initially, a *participation analysis* was carried out to identify relevant actors in the field and to define the role they would play in different phases of the project. The intensity of participation was dependent on this role (e.g. the will to become informed, deliver data, interpret data, evaluate transformation strategies, decide and implement). The participation analysis helped to identify an inner and an outer circle of actors to be integrated with different degrees of intensity. The analysis also

3 More information in Schäfer 2007 and on http://www.regionalerwohlstand.de.

made it clear that the need to integrate certain actors varied in different phases of the project.

The project team decided to use varied participation methods to meet specific objectives:

– Methods that guarantee a close and continuous exchange with the practitioners would allow the integration of popular knowledge.

– Target-specific communication and discussion of the results would be the basis for a continuous exchange.

– Participation of practitioners in the research process would allow them to contribute to decisions.

Collaboration in best practice projects would enable processes to be designed on the basis of scientific results and some of the recommendations to be put into practice.

Table one gives a short overview of the methods and their functions.

Participation Instruments	Function and characteristics
Advisory board with practitioners	Continuous exchange with actors in the field who are highly motivated Ensuring practical relevance of the research questions and the transformational knowledge produced Obtaining additional information about the structure of the sector
Workshops	Exchange with a greater or smaller number of people depending on the phase of the project: a) discussion of the central hypotheses of the project with 35 actors from the field b) participatory reduction of the indicator set with seven representatives from organic associations and three entrepreneurs c) discussion of results with actors from the institutional and regional environment Function: integration of popular knowledge relevant to the specific needs of the project
Participation in formal and informal meetings	Create trust and show interest in the discussions in the field Obtain additional information for the generation of systemic knowledge
Communication of research questions and results in non-scientific publications and events	Public information leaflets and newsletters Publication in journals of practical relevance for the sector Participation in fairs (Grüne Woche, Biofach) Design of an exhibition and a public information booklet Function: to ensure transfer of knowledge generated into the sector

Cooperation in best practice projects	Cooperation in the design of a communication campaign based on the research results Support in developing a code of conduct: "Regional&fair" Charter Cooperation in the initiation of a Leader project supporting direct marketing firms Function: Exemplary realisation of the recommendations The research project was "forced" to link its research questions and results closer to "real life problems"

Table 1 – Participation instruments and their functions in the project "Regional Wealth Reconsidered"

Advisory board with practitioners

At the outset of the project, an advisory board with representatives of the organic sector and associated institutions and organisations was set up (a pool of 18 people). The intention was that the project would be continuously accompanied by a group of people who were highly motivated and ready to offer constructive input. Regular meetings were intended to ensure the practical relevance of the research questions, the transfer of additional information on the structure and the perspectives of the sector as well as feedback on the transformation strategies. In a further stage of the project, the initiation of transformation projects within this group was planned. Further, the members of the advisory board would serve as multipliers spreading information on the project throughout the sector. At the same time, we would be regularly informed about important events and changes in the organic sector.

The advisory board was consulted during all stages of the project, beginning with discussion of the research questions, exchanges on normative premises (the understanding of quality of life) and perspectives of the sector, as well as obtaining feedback on the research design, the central hypotheses of the project and preliminary results. During the meetings, which took place two or three times a year, innovative methods (e.g. fantasy trip, q-method) were applied.

Results: In the first phases of the project, expectations of the role of the advisory board were fulfilled. A lot of information on the sector and constructive feedback concerning the research questions, the normative premises and the research design were obtained. The rather close contact with people from the sector was also very helpful in providing opportunities for further contact with other people from the field.

Problems: Since the members were often very busy, it was actually difficult to convene meetings with more than 10 participants. Due to the irregularity of the board members' participation, it was not easy to guarantee a consistent level of information. During the last year, the frequency of the meetings was reduced. It seemed that for the development of specific transformation projects some of the group could be consulted, but the group as a whole was not always the appropriate forum.

Optimisation potential: By integrating some of the actors as early as in the phase of problem definition (application phase), it might have been easier to collaborate as "equal partners" instead of having the practitioners act as commentators or observers.

Participatory reduction of the indicator set

Within the RWR project, a set of indicators was developed to analyse the contribution of the organic sector towards quality of life and sustainable development. After discussions about the understanding of quality of life in the advisory board, and integration of feedback on the hypotheses of the project, the research team developed an analytical framework on the basis of current conceptions of sustainability and quality of life. Based on extensive literature research, this framework was made operational using criteria and a set of 150 indicators. This top-down process of defining indicators was combined with a participatory "bottom-up" process to select the most appropriate indicators in the given regional and industrial context.

Seven regional representatives from associations of the organic agriculture and food sector and three managers of regional firms (organic agriculture, food processing and trade) were asked to vote on which of the suggested indicators would be most relevant or best suited to describing the contributions of the sector toward quality of life and sustainability. Then, in a workshop, we discussed why some indicators were voted more relevant than others.

Results: This participatory step allowed the set to be reduced to 70 indicators, without substantial loss of completeness. Incorporating the knowledge of those analysed was very helpful for the evaluation and selection of criteria and indicators that – in our opinion – cannot be done in an "objective" manner. Furthermore, the discussions about defining the indicators facilitated the exchange of opinions about normative premises and project goals, making them appear more concrete. Additionally, participation in selecting the relevant indicators created greater acceptance for the survey amongst the sector representatives. After the workshop, all of them were willing to motivate their members to participate in the survey, publishing a short announcement in their associations' journals.

The expectations we had of this participatory step were completely fulfilled.

Problems: For this participatory step the structure of the group had to be carefully considered in order to guarantee balanced comments. Additionally, questions arose about the importance of the practitioners' evaluation in relation to that of the scientists.

Participation in formal and informal meetings of the sector

Members of the research team regularly participated in formal or informal events of the sector, such as meetings of the associations, festivities and fairs. This participation signalled interest in current events in the sector and allowed additional internal information to be obtained. It also helped to consolidate relationships and build trust. Due to this kind of commitment, members of the research team have been invited to participate in an additional project analysing perspectives for the organic sector in East Germany.

Our expectations on participation in formal and informal events of the sector were completely fulfilled. There is, however, always a conflict of time commitment; and deciding between tasks that are more clearly of a scientific nature and those that are aimed at consolidating relationships with the actors in the relevant field is difficult.

Preconditions for this type of exchange with practitioners include: personal engagement or being interested in this field, the use of comprehensible (non-scientific) language and the ability to translate "everyday problems" into scientific terms. To be taken seriously by the actors in the field, it is necessary that they be continually informed.

Effects of the research project on the organic sector

The research project was conducted during a phase of intense reflection within the organic sector about its role in society and thus has been able to contribute to this debate with its research results. The current discussions are taking place against a background of the growth of the organic market and the processes of differentiation this entails. In addition to the idealistic pioneers who have always been strongly linked to social and environmental movements, new actors with a stronger economic motivation to take part in a growing market have been entering the sector. This process has forced the pioneer firms to re-position themselves on the market by highlighting the quality of their products and their societal engagement, which goes beyond producing organic products. The results of the research project have acted as an impetus to debates about future perspectives and further development strategies within the organic sector. In one of the best-practice projects it was collaborating in, the project supported the actors in defining a set of criteria regarding social and ecological qualities that go beyond existing organic regulations. Communicating this additional societal engagement has become part of an "organic plus" marketing strategy, which differentiates these firms from others in the organic sector.

A public exhibition and public information had an important communicative function within this discussion process, as practitioners at different levels (businesses, political parties, associations, etc.) used them within the context of their own debates. Additionally, the researchers for the project were often requested to provide consultation and presentations.

Besides participating in and facilitating the debates within the organic sector, the research project was able to integrate its results into the debate about future perspectives and appropriate funding strategies for rural areas in Europe, questions that are being very intensely discussed within the context of EU agrarian policy at national and regional levels. The project contributed to this debate by highlighting the role (organic) agriculture is playing in society, beyond the mere production of food.

Summing up, the project can be said to have succeeded in developing a set of recommendations that were very well embedded into the regional context of the organic food sector. From our point of view, the specific quality of transdisciplinary research is its continual alternation between scientific (theoretical) and external (real world) perspectives and reflection on the differing demands of these two fields of activity. This type of research yields results and solutions that go beyond the usual range of the two types of actors involved, because they are both closely linked to "real problems" and, at the same time, based on scientific evidence. The role of science in these processes is to stimulate debate with innovative recommendations and solutions; the task of putting these recommendations into practice mainly has to be carried out by the actors themselves. Incentives provided by transdisciplinary projects will, therefore, have a rather indirect effect.

Discussion

The experience gained in some of the Social-Ecological Research Programme projects makes it possible to further systematise experiences concerning the relation between the aims of integrating popular knowledge and the methods used.

In the project "Regional Wealth Reconsidered", we followed a participative approach, which assumed that including the problem definition, interests and recommendations of the organic sector actors as well as of their context (consumers, political actors, etc.) is a precondition for being able to develop sustainability solutions of practical relevance

(Bergmann *et al.*, 2005; Boeckmann *et al.*, 2005; Schäfer, 2007). The research project therefore has one foot in the scientific community – because the results have to be arrived at according to scientific standards – and the other one in the lifeworld, where the results are meant to contribute towards further development. Conflicting demands between scientific and practical requirements are therefore deliberately taken into account in the research process, because a special potential for innovative solutions is assumed to be found in this field of interaction (Loibl, 2004; Pfriem *et al.*, 2006). Being part of two contexts with different logics can result in conflicts between the demand for scientific objectivity and the development of applicable solutions. The difficulties entailed (which are also inherent to disciplinary research) need to be reflected upon openly and systematically in order to result in synergies for the research process (Nölting *et al.*, 2004).

To be able to evaluate the strengths and weaknesses of a transdisciplinary sustainability research project, it is necessary to reflect upon its design and the approaches chosen on a more abstract level. Pohl and Hirsch Hadorn (2007) identify four distinct design principles to meet the specific demands of transdisciplinary research. These principles are reflected in our research design. By *mapping* the relevant actors and their interests, the complexity of lifeworld problems was reduced (instrument advisory board and participative selection of indicators). *Embedding* the research project in the regional context and the organic food sector as well as in regional development processes improved the applicability of the solutions developed (instrument advisory board, public exhibition/brochure, best practice projects). *Open encounters* with different actors helped to integrate different perspectives and normative positions from various disciplines and societal groups. Last but not least, the research process was *recursive*, since we used instruments such as the advisory board, which allowed a long-term continuous exchange between science and lifeworld. These discussions and the resulting constructive inputs had the function of "reflective loops".

The experience of the project shows that there is not one single method for transdisciplinary research. The principles point to the importance of different approaches, but the mix of methods used needs to be adapted to the specific problem dealt with and the actors involved.

Comparing the project goals and the instruments used with the general design principles of transdisciplinary research, it becomes clear that the variety of methodological instruments used was appropriate for achieving the project goals, on the one hand, and sufficiently respected the general design principles, on the other.

Project goals of the transdisciplinary approach	Methodological instruments	Design principles of transdisciplinary research
a) Integration of lifeworld knowledge	Advisory board of practitioners Other instruments: Workshops with practitioners Regular participation by the researchers in meetings and events of the organic sector	Mapping Contextualisation Open encounters Reflexivity
b) Continuous exchange with practitioners	Brochure Exhibition Other instruments: Flyer Website Newsletter Publications Consulting on boards of the organic sector	Contextualisation Open Encounters
c) Options for contributing towards decision-making within the project	Participative selection of indicators More instruments: Workshops with practitioners	Mapping Open encounters
d) Putting recommendations into practice	Planning and carrying out joint best practice projects	Contextualisation Open encounters

Table 2 – Project goals, methodological instruments and principles of transdisciplinary research

This systematic comparison highlights the central importance of the advisory board of practitioners, an instrument that contributes to the four design principles of transdisciplinary research. Besides integrating lifeworld knowledge and guaranteeing continuous exchange, it also offered possibilities to contribute towards decision-making and was crucial in preparing the best-practice projects. This instrument proved to be an innovative element, transferrable to other research projects.

Despite the central role of the advisory board, the experience gained shows that the complex demands of transdisciplinary research cannot be fulfilled by one method alone. Table 2 shows how the different instruments complement each other; for example, for exchange with practitioners, several target-group-specific instruments were necessary.

With reference to Bergmann *et al.* (2005), the participatory instruments used for the project contributed towards provision of knowledge about the field of interest and towards the integration of normative and transformative knowledge into the sector. Problem definition by the project researchers was carried out in a "precautionary sense" (Bergmann *et al.*, 2005), and interest in the project shown by the practitioners proved that the topic chosen was of practical relevance.

A special characteristic of the project was explicit reference to the normative concept of sustainable development. The project positioned itself clearly when it comes to the applied understanding of sustainability and treated questions of normative positioning in a transparent manner. Some actors associated the project with certain concepts and political positions due to this normative contextualisation. Others, however, used this normative debate as an incentive for reflection and discussion. Dealing with different interests and values proved to be very fruitful for accomplishing the project goals of stimulating discussion about future perspectives for the organic sector and its role in society.

Conclusion

The experience gained in the above projects of the Social-Ecological Research Programme, and more specifically within one of these projects, highlights how the methods used in transdisciplinary sustainability research need to be chosen in such a way as to link to the specific goals being pursued via participation and the type(s) of knowledge production they are aiming for. In projects aimed at producing transformational knowledge, or at intervening in development processes, exchange with practitioners is generally more intense than in projects aimed at deepening systemic or normative knowledge through the integration of lifeworld knowledge. To be able to fulfil the general design principles of transdisciplinary research – contextualisation, embedding in the context, approaching each other openly and reflexivity – an appropriate mix of instruments is necessary and needs to be adapted to the specific design and objectives of the research project.

References

Balzer I., Wächter M., Sozial-ökologische Forschung. *Ergebnisse der Sondierungsprojekte aus dem BMBF-Förderschwerpunkt*, Munich, Oekom, Gesellschaft für ökologische Kommunikation, 2002.

Baranek E., Fischer C., Walk H., *Partizipation und Nachhaltigkeit. Reflektionen über Zusammenhänge und Vereinbarkeiten*, *Discussion paper*, Berlin, Zentrum Technik und Gesellschaft der TU Berlin, 2005.

Becker H., *Ostdeutsche Dörfer im Aufbruch. Fallstudien ländlicher Entwicklung*, Bonn, FAA, Schriftenreihe der Forschungsgesellschaft für Agrarpolitik und Agrarsoziologie, 2000.

Bergmann M., Brohmann B. *et al.*, *Qualitätskriterien transdisziplinärer Forschung. Ein Leitfaden für die formative Evaluation von Forschungsprojekten*, Frankfurt am Main, Studientexte des Instituts für sozial-ökologische Forschung (ISOE), 2005.

Boeckmann T., Schäfer M., *Anforderungen an das Projektmanagement in der transdisziplinären Forschung* (unpublished manuscript), 2004.

Boeckmann T., Nölting B., "Ökologische Land- und Ernährungswirtschaft in Berlin und Brandenburg", *Ländlicher Raum*, 56, 1, 2005, pp. 16-21.

Brand K.W., "Nachhaltigkeitsforschung. Besonderheiten, Probleme und Erfordernisse eines neuen Forschungstyps", in Brand K.W. (ed.), *Nachhaltige Entwicklung und Transdisziplinarität. Besonderheiten, Probleme und Erfordernisse der Nachhaltigkeitsforschung*, Berlin, Analytica Angewandete Umweltforschung, 16, 2000, pp. 9-28.

CASS/ProClim, *Forschung zu Nachhaltigkeit und Globalem Wandel. Wissenschaftspolitische Visionen der Schweizer Forschenden*, Bern, Proclim, 1997.

Functowicz S., O'Connor M. (eds.), "Science for Sustainable Development", *Special Issue of the International Journal of Sustainable Development*, 2, 3, 1999.

Functowicz S., Ravetz J., "Science for the Post-Normal Age", *Futures*, 25, 1993, pp. 735-755.

Gibbons M., Limoges C. *et al.*, *The New Production of Knowledge. The Dynamics of Science and Research in Contemporary Societies*, London, Sage, 1994.

Gibbons M., Novotny H. *et al.*, "The Potential of Transdisciplinarity", in Klein J. *et al.* (ed.), *Transdisciplinarity: Joint Problem Solving among Science, Technology and Society*, Basel, Springer, 2000.

Grunwald A., "Transdisziplinäre Umweltforschung: Methodische Probleme der Qualitätssicherung", *TA-Datenbank Nachrichten*, 8, 3-4, 1999, pp. 32-39.

Häberli R., Grossenbacher-Mansuy W., "Transdisziplinarität zwischen Förderung und Überforderung. Erkenntnisse aus dem SPP Umwelt", *GAIA*, 7, 3, 1998, pp. 196-213.

Hirch H. Maier G., Wölfing K.S., *Transdisziplinäre Forschung in Aktion*, SPPU, Zurich, Schwerpunktprogramm Umwelt, 2002.

Loibl M.C., "Spannungen in heterogenen Forschungsteams. Prioritätenkonflitke nach Wissenschaftskulturen, institutionellen Zugehörigkeiten und Gender", in Strübing J., Schelz-Schaeffer I., Meister M., Gläser J. (eds.), *Kooperation im Niemandsland. Neue Perspektiven auf Zusammenarbeit in Wissenschaft und Technik*, Opladen, Leske & Budrich, 2004, pp. 231-247.

Luley H., Schramm E., *Optionen in der Nachhaltigkeitsforschung, Materialien Soziale Ökologie*, Institut für Sozial-Ökologische Forschung, Frankfurt am Main, 2003.

Maier Begré S., Hirsch Hadorn G., "Options & Restrictions: a Heuristic Tool in Transdisciplinary Research for an Effective Implementation of Sustainable Practices", in Biermann F., Campe S., Jacob K. (eds.), *Proceedings of the*

2002 Berlin Conference on the Human Dimensions of Global Environmental Change Knowledge for the Sustainability Transition, the Challenge for Social Science, Global Governance Project, Amsterdam, Berlin, Potsdam & Oldenburg, 2004, pp. 322-336.

Nicolini M., *Sprache, Wissenschaft, Wirklichkeit. Zum Sprachgebrauch in inter- und transdisziplinärer Forschung*, Klagenfurt, Vienna, Bundesministerium für Bildung, Wissenschaft und Kultur, 2001.

Nölting B., Voß J.P., Hayn D., "Nachhaltigkeitsforschung - jenseits von Disziplinierung und 'anything goes'", *GAIA*, 13, 4, 2004, pp. 254-261.

Pfriem R., Raabe T., Spiller A., "Von den Vorteilen und Risiken transdisziplinärer Forschung", in Pfriem R., Raabe T., Spiller A. (eds), *OSSENA – Das Unternehmen nachhaltige Ernährungskultur*, 35, Marburg, Metropolis-Verlag, 2006, pp. 473-484.

Pohl C., Hirsch-Hadorn G., *Principles for Designing Transdisciplinary Research. Proposed by the Swiss Academies of Arts and Sciences*, Munich, Oekom, 2007.

Schäfer M. (ed.), *Zukunftsfähiger Wohlstand – Der Beitrag der ökologischen Land- und Ernährungswirtschaft zu Lebensqualität und nachhaltiger Entwicklung*, Marburg, Metropolis, 2007.

Schophaus M., Schön S., Dienel H.L. (eds.), *Transdisziplinäres Kooperationsmanagement. Neue Wege in der Zusammenarbeit zwischen Wissenschaft und Gesellschaft*, Munich, Oekom, 2004.

Siebenhüner B., Siebenhüner B., "Social Learning and Sustainability Science: Which Role Can Stakeholder and Participation Play?", in Biermann F., Campe S., Jacob K. (eds.), *Proceedings of the 2002 Berlin Conference on the Human Dimensions of Global Environmental Change "Knowledge for the Sustainable Transition. The Challenge for Social Science", Global Governance Project Amsterdam*, Berlin, Postdam and Oldenburg, 2004, pp. 76-86.

Tress B., Tress G. *et al.*, *Interdisciplinary and Transdisciplinary Landscape Studies: Potentials and Limitations, 2*, Wageningen, DELTA (DELTA Series), 2003.

Wynne B., "Unruly Technology. Practical Rules, Impractical Discourses and Public Understanding", *Social Studies of Science*, 18, 1998, pp. 147-167.

Assembling Different Forms of Knowledge for Participative Water Management

Insights from the Concert'eau Game

Audrey RICHARD-FERROUDJI and Olivier BARRETEAU

Researchers, IRSTEA, Montpellier, France

In the environmental field, water management provides examples of the current move towards participative democracy. During the past 45 years water policy has moved from a technocratic sector-based and centralised form of management to one that is more local, integrated and participative. Water policy also increasingly considers the river basin as the relevant area for water management. French water policy promotes concerted river basin management settings and provides incentives for a move towards collective decision processes. However, river basin management settings place different forms of knowledge in contact with each other, raising the question of how to combine this plurality within the decision-making process. For example, an advisory committee dealing with the flooding issue in the Orb Valley (Richard-Ferroudji, 2003) featured participants discussing flood management action. While talking about flooded campsites, a campsite manager was asking for public financing of dikes to protect camp sites:

> "We are always the ones asked to pay," said the manager. "Supermarkets benefit from tourism but pay nothing. Campsites provide the region with tourism and make it wealthier [...].
>
> An elected representative said: "The amount involved is nothing for this area. Your problem is you lack political support [...]. You need somebody to represent you with local government.
>
> The campsite manager said: "[...] Nobody cares if I go bankrupt."
>
> The elected representative/meeting chairman replied: "I understand your position. It is natural. You are raising the general issue of the civil service trying to grab jurisdiction over questions that decentralisation gave to elected representatives. [...] This is the first time that I've seen you speak out like this but it's all too much for you."

Several arguments were being raised here: on the level of value, with justification of public support through market arguments; on the strategic level, concerning distribution of power; and on a more emotional level with the indifference to the campsite manager's bankruptcy. The meeting also dealt with French risk policies, laws, a decision-maker's lack of legitimacy, as well as dike efficiency, both according to an engineering study and according to participant experiences. This example illustrates how scientific, lay and diverse cultural forms of knowledge interact in collective decision processes. How do participants handle such different forms of knowledge? How do they reach decisions? This paper focuses on the assembling of different knowledge and "goods" for collective decision-making. It first presents the theoretical background, then the case study with the methods used, including a game. Finally, it reports on preliminary results of its experimentation and discusses these results.

Analytical approach: assembling several forms of knowledge

River basin management programmes propose different spaces for collective debates, such as advisory committees or river basin committees. These spaces can be described as "hybrid fora": major deliberating mechanisms to manage controversies over scientific and technological innovations (Callon, Lascoumes and Barthes, 2001). According to these authors, a hybrid forum must be a space where those taking part can explore options and learn together, a process in which the identity of participants may change or be built up over time. Popular knowledge is not discredited and considered illegitimate; on the contrary, it is respected and taken into consideration. The aim is to open up free debate among all of the parties affected (including scientists, industrial corporations, institutions, associations and the public), so all opinions can be heard and respected. To make the assembling of knowledge possible, a setting should therefore allow diversity to be defined by the participants themselves; otherwise this will happen outside the debate situation. Even non-human beings (such as fish or floods) are aligned with human beings in actor-networks through processes of "translation". The aim is that the result of the debate can be taken into account in the decision-making process, in a manner that is transparent from the outset. In fact, a hybrid forum manages controversy, leaning on the principles of symmetry and publicness. Debate is framed such that any piece of knowledge can be proposed, discussed and potentially included in its hybrid outcome. This issue of hybridisation or translation has given birth to a wide range of mediation tools and methods: focus groups, public debates, brainstorming, stakeholder assessments, participatory modelling, etc. (Selin and Chavez, 1995; Cockes and Ive, 1996; Babin, Bertrand *et al.*, 1999; Kraft and Johnson, 1999).

All these tools and methods are aimed at combining various sources of knowledge. They bring together an older tradition in the field of support to public policies through the use of models, be they computer based, maps or mere verbal metaphors (Saunders-Newton and Scott, 2001). While these were focused toward a single client, collaborative settings make the issue of interfacing the tools with their public of use more difficult. Recent developments in computer sciences have provided models, such as Agent-Based Simulation models (Ferber, 1999; Moss and Davidsson, 2001), which are able to gather heterogeneous sources of knowledge and can be used in interactive settings (Bousquet, Barreteau *et al.*, 2002). In recent years these mediation support tools have flourished. While they sometimes differ only in the terms used by their authors, they nevertheless often differ in the artefacts deployed to facilitate the assembling process, their institutionalisation and their scale of intervention (Dzizdzicki, 2001). They are all aimed at achieving an agreement among participants, which may involve organising trade-offs among interests or very long-term "patrimonial" objectives (Babin, Bertrand *et al.*, 1999).

How does hybridising occur in practice, according to the "goods engaged"? The issue of better understanding the conditions for a real debate, and assembling the diversity of views available, accompanies the recent spread of mediation support tools. To tackle this issue, we focus on the hybridisation process in context. In the introductory example, participants have to simultaneously take in a participant's bankruptcy, EU policy for risk management and criticism of "public services". How can they deal with all these concerns in the same framework? We analyse how people act in hybrid fora, using the framework of "regimes of engagement", which links cognition and action (Thévenot, 2000). We observe people acting in collective discussions, relying on "pragmatic regime" categories (Thévenot, 1999). This framework aims "to account not only for the movements of an actor but also for the way his or her environment responds to him or her and the ways that he or she reacts to these responses". "Regimes of engagement" link "the reality and the good engaged". A person's good is understood as "what is good" for him or her. Using this framework entails considering people as autonomous beings engaged in interactions, with moral capacities (values and principles), strategic capacities (interests and intentions) and feelings and emotions. The kind of good that governs their engagement in a real situation varies from personal and local convenience to collective and legitimate conventions.

Justifying water management approaches

Firstly, hybrid fora feature participants justify the legitimacy of arguments within the democratic polity. Actors demonstrate the situational appropriateness of their criteria of evaluation and find material proof that their arguments are grounded. For example, the camp site manager argues for funding the camp sites' protection because they make the region wealthier. He underlines the importance of tourism infrastructure for the river basin's economic development. At this stage, he is acting in the discussion in a "regime of justification" (Boltanski and Thévenot, 1999 and 2006). In this regime, people offer many different descriptions of the situation, relying on various conventional types of information. Boltanski and Thévenot identify "orders of worth" which constitute "common forms of public evaluation and which are grounded on the same grammar of the common good". Thévenot (2000) studies a road and tunnel project controversy and describes the plurality of "worthy roads". The road can be described as "a highway of market worth" which is opening up landlocked areas to market competition. The project can also be justified as efficient infrastructure. The road can also be seen as a "customary way of integrating locals" or "a famous scenic route". But opponents can argue that it is an environmental scar. In our introductory example, the campsite manager justifies his claim for market reasons.

This regime of justification leads to a first definition of assembling plurality and forms of knowledge, as deciding which arguments are more legitimate. For example, the campsite manager's claims can be considered more or less legitimate than the environmental protection issue. The arguments' legitimacy depends on the context and the participants. Thévenot, Moody and Lafaye (2000) compare arguments and modes of justification in the above road project in France and a river dam project in the United States. They show, for example, that market evaluations were common in the United States and were often combined with other sorts of evaluations such as civic arguments. Market arguments for the project were also used in France, but primarily came from Brussels and were seldom endorsed at the local level. Moreover, civic arguments were not combined with market arguments, but with planning arguments.

Defending interests in river basin management settings

The analysis of how knowledge is assembled within the justification regime takes into account the participants' moral capacities for defending common goods. Yet other capacities are deployed in discussions, such as strategic capacity. People implement intentions and use arguments to do so. Thus we should analyse the assembling of knowledge by connecting the actors' vision of "community" to the actors' interests (Moody and Thévenot, 2000). Besides justification, people enter discussions to defend their interests in a pragmatic regime of "regular planned action". In dialogue settings, participants can be considered as "stakeholders" who implement intentions and consider objects as functional. The campsite manager who asked for help to avoid bankruptcy is using the advisory committee to protect his interests. Elected representatives defend their power against "civil servants". Salles *et al.* (1999) analyse farmers' strategies within water management settings. They study the integration of environmental norms in farming practices. They underline farmers' "strategic appropriation" of "agri-environmental" measures without real changes in practice. They identify four farmer strategies for entering the discussion in a strategic regime.

This regime of "regular planned action" leads to a second definition of the assembling of plurality: negotiating. Many authors have analysed negotiations for water management issues, yielding proposals for designing negotiation processes (Allain and Emerit, 2003; Mermet, 1998). The campsite owner quoted above is trying to adopt such a negotiated approach. His arguments are aimed at mixing interests, seeking a win-win solution through the assumption that an overarching shared interest for actors is the presence of tourists in the area. He could say: "I bear the physical cost of floods, so could other people please bear the financial cost?" However, being in a repeated interaction series with the other participants, and notably the meeting chairman, their past relations are invited into this negotiation process (last sentence of the extract from the discussion above). The gentle reminders from the meeting chairman also help to reaffirm all the power relations, as well as any debts the camp owner might have.

This second point of view on the assembling of knowledge, then, paves the way for making use of the interactions among varied interests and enlarging the solution space in order to seek some win-win solutions.

Assembling knowledge without excluding knowledge difficult to express in public arenas

But there is more to the campsite manager's claim during the committee meeting than a mere attempt to defend his interests or justify the legitimacy of his claim. There is emotion. He is voicing his strong ties with his campsite, the long-standing relationship to this place, built through a lifetime of work. This kind of relationship with surroundings or people requires a different way of qualifying actions, other than seeking to justify or achieve a goal. Thévenot proposes a third "pragmatic regime" of "familiarity" which depends on idiosyncratic linkages with a customised environment. It has to do with "perceptual and kinaesthetic clues about familiar and customised paths through local environments which involve modifying the surroundings, as well as the habits of the human body [...]. The proper language to offer accounts of what happens is far from the formal statements offering justifications. It is highly indexical and gestural" (Thévenot, 2000). How do people cope with such relationships in dialogue settings? They have first to cope with their attachments to the environment while participating, and then make people and things equivalent and general.

Doidy (2003) studies the difficulty of voicing this engagement with the environment in collective decision-making. He describes a "proximity" kind of knowledge through the relation between fishermen and the environment termed "water sense". He underlines the difficulties encountered when trying to "value" this "proximity" knowledge and relationship in public arenas. A farmer also points to this tension of getting others to understand his relationship to the environment, of rendering it general or equivalent to that of other people: "You can't know what farming is if you haven't experienced it." "Familiar engagement" is undermined by the process of making things public. The campsite manager fails to voice his personal relationship. We feel emotions but words do not come and the elected representative immediately moves the debate along towards collective issues. In our introductory example, the elected representative is sympathetic to the campsite manager's problems and responds by allowing him to make his claim general and available to a public assembly. But sometimes claims that fail to be shaped for the public can lead to tensions or even violence because certain voices cannot make themselves heard.

What place can be given to such personal ties and the proximity kind of knowledge? Is this only an individual interest, which has no place with respect to the general interest? Or is this a private issue that should not be treated in public places? This knowledge cannot only be considered as hidden information that the setting should reveal. The division between private and public does not hold, and shifts between forms of proximity and public stance should be considered seriously. While personal ties should not guide public decision-making, they should not be totally ignored either. From our analytical point of view, what happens in participative settings has to be looked at taking these personal ties into account, be they explicitly introduced or merely in the context. Actors cannot leave all their personal "equipment" at the door of the arena. They will thus use it in the collective decision-making process anyway.

Doidy calls for reflection on how environmental management settings can integrate this "proximity" type of knowledge and take it into account in the assembling plurality process. Many dialogue settings only accept knowledge shaped for the public, and above all scientific knowledge. Even if hybrid fora accept different categories of knowledge, these types of knowledge have to be understandable and thus translated for the public. The question of allowing "proximity knowledge" and its translation still remains. We make the hypothesis that water management settings should allow personal knowledge and ties to enter the process and could even benefit from accommodating it. They should accept people with all their facets, and not only individuals defined as public, autonomous and responsible.

Shifting between different "pragmatic regimes"

Following a pragmatic stance, this paper studies how "participants" combine different forms of knowledge. We will focus neither on which arguments are considered more legitimate and fair, nor on how people negotiate and on power distribution. Here, we specifically focus our attention on the assembling of plural cognitive forms, from the most personal to the public. Finally, we consider various forms of knowledge at three levels linked to the three pragmatic regimes presented: "familiarity", "regular planned action" and "justification". Are participants talking about what is for the common good of the river basin? Are they talking about different interests and negotiating? Or are they trying to voice something more personal? How do people deal with their attachments to their environment? How best to combine values, interests and ties for collective water management? Where are difficulties involved in the assembling? And finally, how should a participative setting cope with this assembling of knowledge? Following the "pragmatic regimes" framework, we have observed people's shifts between different "pragmatic regimes of engagement" in various existing or experimental settings. We focus on the abilities required to shift from one pragmatic orientation to another, depending on arrangements specific to the situation.

Summary of the case: Lentilla and Llech river basin

The case under study[1] here is the Lentilla and Llech river basin's collective management. The Lentilla and the Llech are two Mediterranean rivers in the south of France in the Catalan area. The river receives its water from Canigou, a famous Pyrenean mountain peak. The river basin covers 9000 ha. The climate is Mediterranean, characterised by acute scarcity of water during summer. In this season, demand for water

[1] This description is based on 16 interviews and the observation of one meeting of the collective project management committee. Translations of French quotes from the case are the authors'.

resources is high and can lead to competition among different water uses, including farming, fishing and potable water.

Water sharing among stakeholders

During dry summers, farmers divert all of the water from the Lentilla to irrigation channels in order to irrigate their fields (mainly peach trees). Competition between water for drinking and water for farming has existed since the 1960s and competition with fishermen became conflictual a few years ago. For many actors, "the challenge is thus to ensure water during low water periods, making it possible to answer the whole of the basin users' needs".[2] The issue is then expressed in this context, through a choice between water for farmers and water for fishermen.

Many water management participants agree on a plan to pump water dammed down river, to supply water to the river and satisfy all the demands. However, funding bodies (the French water agency and municipalities) make their funding dependent on a real collective and integrated management strategy and on the definition of a "low-water management plan".[3] The first condition makes it necessary to consider all water issues, such as improving water quality and not only the summer scarcity issue. The second condition makes a public agreement necessary, whereas the existing political modalities of management are mainly of a community type and lean on interpersonal arrangements. When public meetings occur, their only aim is to announce public authorities' or elected representatives' decisions. The debate is hijacked into "partly private" settings where participants know each other and know they can agree. There is no public discussion. The challenge is to craft a water management setting to accommodate public debate about water sharing.

Here we see the necessity of assembling the multiple views and relations to the river basin. Actors view this plurality and shape it for the public through interest categories: for example, farming versus fishing. In the observed meeting, people were mainly part of a "regular planned action" regime, and sometimes one of "justification". As the predominance of agricultural water use is being challenged by environmental issues and by the increased use of water for leisure activities, farmers have adopted a strategic attitude to maintain some power over water management and to be included among the planning group leaders

2 Authors' translation. Source: Feraud J., Chambre d'agriculture des Pyrénées-Orientales, "Aménagement hydraulique du bassin de la Lentilla", 2003.

3 "Protocole de gestion des étiages".

(Salles *et al.*, 1999). But the hierarchy of uses is never discussed publicly.

Approaching water issues through interest categories seems to be a dead end, as this viewpoint separates the categories, whereas this is not what the users themselves do. An elected representative told us: "I conceal the farming issue. I speak less about it because I'm worried. It affects me." A teacher, and president of a canal association, also voiced her anxiety about the future:

> We own two uncultivated fields. We don't really know what we're going to do with them. I'd like to see trees on them again, as there were in the past. It's a pity to see things disappearing like that. But it could be argued that it's hard to make a living from it. I don't know how some people manage to make a living from it. That's why water for cultivated fields is a priority.

This person is not comparing farming with fishing, but rather heritage preservation with economic optimisation. She is comparing two criteria of evaluation. In the same way, we have attempted to identify water management "logics of evaluation" and to understand the assembling of knowledge through the "justification" approach, transversal to the interest approach. In the following sections we describe the valley, as seen from four viewpoints about the common good, and present illustrative statements.

What is good for the Lentilla and Llech river basins?

The river is the heritage of the inhabitants of the "barony". Traditions and "patrimony" must be respected: "These century-old canals serve [...] patrimonial functions. They are witnesses to history and water culture." Very ancient irrigation canals cover the Lentilla and the Llech valleys. The canal in the downstream Lentilla Valley was created in 1282 and covers a territory still called "the barony", referring to the political organisation that existed at that time. Rules for sharing water were locally defined. A long tradition of collective water sharing and management exists (Broc *et al.*, 1992; Ruf, 2000). Farmers' associations cooperate to manage the water, as well as water scarcity when it occurs. Conflicts over water have always existed and are solved by informal arrangement among representatives of the different users. For example, farmers and elected representatives phone each other when there is a lack of water. Sometimes the elected representative is a farmer himself. Many inhabitants also have family links and rural or farming origins. Water management based on strong social links has proved efficient in the past, but today it is severely stretched as it is being required to integrate newly-affected people and newcomers.

– The river is a resource whose uses must be technically and economically optimised: "We have to recognise hydro-systems as natural capital generating services [...]. It is necessary to make the value of the services visible, whether they are real or potential, then to compare them with the investment required to protect them" (Agence de l'eau RMC, 2003). From a technical point of view, water management must be optimised. From the economic point of view, the valley has long been a farming valley. Since the creation of the canal and up until a few years ago, farming developed on the "barony", while irrigation efficiency increased. Today, farmers are facing quality requirements and European farming competition. Tourism and leisure water uses are increasing and now compete with farming. Fields are sold to build houses. The farming population is decreasing. Pumping dammed water could facilitate economic development and needs to be evaluated both from a technical and from an economic point of view. The river basin needs economic development. Water is an economic good and represents major capital for the river basin through tourism, leisure and agriculture.

– The river is a common good that must be managed with respect to the public interest: "Water forms part of the national heritage [...] water use is everyone's right within the framework of the laws and regulations [...]."[4] An association of municipalities has led the pumping project since January 2005, to guarantee the respect of the public interest, whereas previously the project was led by fishermen and farmers. Solidarity and civic equality are major issues. The river is a space that should be shared by all. River basin management should respect water law and no longer wave the oldest rights like a banner. It should follow public opinion, of the inhabitants of the river basin and downstream. Moreover, inhabitants should be citizens first, before being consumers or the valley's children. Canals are famous collective symbols of the Pyrénées-Orientales region (the French part of Catalunya).

The river is alive and man should live in harmony with it by protecting the natural equilibrium and environmental beauty: "The river shouldn't be mistaken for a duck pond. Here you are at the heart of nature. Nature needs to be respected. The fish is a noble creature." The upstream part of the river is narrow and cliff-lined. It acts as a sanctuary for animals and fish, among them trout. The landscape is beautiful and wild, containing a Natura 2000 protection area. Downstream the river has been modified by human activity. Fish are endangered by pollution and water scarcity. The aquatic environment and non-human beings should be respected. Diversity needs to be preserved. Pumping water

4 French water law, 1992, article 1

from the dam is "tinkering", "heavily modifying" and "something that is not natural".

Methodology: using simulation games to study the assembling of knowledge in collective decision processes

Our aim was to understand in a given situation how certain actors act in response to specific events and how they try to combine multiple values, interests and personal ties. Given the theoretical framework presented above, we intended to test this framework and analyse behavioural patterns in participative water management settings. An approach based entirely on the observation of real participation settings would yield a thorough analysis of the diversity of sources of argumentation mobilised. However, some of the pragmatic regimes considered do not always appear explicitly. This means that they might be difficult to observe. An approach based only on observation would make it difficult to generate and test how generic and relevant our framework is. Therefore, we tried to vary the inquiry methods (Cheyns, 2005) and observation stance, adding an experimental tool to investigate the assembling of plurality in controlled settings, in order to obtain information and repeat scenarios for the sake of comparison. Besides interviews and meeting observations, we chose a companion modelling approach (Bousquet *et al.*, 2002; Bousquet *et al.*, 1999), implementing the use of a gaming artefact. The companion modelling approach is a framework that specifies a stance when it comes to designing and using models in interaction with stakeholders. Important features of this approach are:

– This is a cyclic process, iterating through the design of virtual worlds to represent the dynamics at stake, the joint exploration of these virtual worlds and the elaboration of the consequences of this exploration for the real world (Etienne, 2009).

– Models are bound to be criticised and modified through their use, or the use of their simulations, with stakeholders. They are tools to obtain information and never an end product of the process. Here we consider a sociological model of collective discussion.

Companion modelling can be used for different purposes: research or collective decision-making support. This separation needs to be considered in the light of a primary objective in the design of a companion modelling process, according to which outcomes should be analysed and assessed. In all cases, the category that is not the primary objective appears to be a side effect of the process that should be taken into account: research in interaction with actors may induce changes in their community while collective decision support can lead to new knowledge on a system itself or on scientific questions. Here we encounter our first research objective since our purpose was to understand the assembling of plurality of "goods" in a collective decision process. Interviews and the observation of meetings led to the design of a representation of a collective decision process where events require interactions concerning water management issues. This model, Concert'eau, is described fully in the next section. Interviews and observations formed the basis for contextualising the model and identifying key issues involved in assembling plurality.

The model was applied in an experimental setting, a role playing game (RPG). RPGs are group settings that determine the roles or behavioural patterns of players, as well as an imaginary context. An RPG is the performance of a roughly defined situation that involves people with given roles (Mucchielli, 1983). Players can stand in another person's shoes and think hard about the roles of the other parties. Players genuinely use an RPG as a social laboratory. It is a way for them to experiment with a variety of ways of positioning themselves in a group with presumably few consequences in the real world (Innes and Boother, 1999). At present, RPGs are used alone as training tools and are also becoming scientific tools. As a group setting, they are suitable for negotiation or collective decision-making (Barreteau, 2003). As training tools, they have already proven to be powerful in stimulating and supporting coherent group change (Tsuchiya, 1998).

Experimental settings are not totally new in the social sciences. They are quite common in economics, where they usually implement a theoretical model to be tested with real economic agents in a very constrained framework. They are also common in management science through the implementation of policy exercises (Toth, 1988). They have been inherited from war games and are aimed at putting participants in a context that could occur, making them react and interact among themselves through a simulation exercise and getting valuable feedback from their collective outcomes for their own practice. From a sociological analysis point of view, the purpose of the game is to provide elements of context to participants and analyse their reactions, with the assumption that these elements of context are the main drivers in the behavioural patterns in the game (Boltanski and Thévenot, 1983). The aim of experimentalists using RPGs for this purpose is not to simulate reality. The game is a way of piecing together a controlled complexity on which to work (Barreteau, 2003). RPGs, therefore, provide a way to take into account the multiplication of decision centres within their setting: the distribution of decision-making processes among all players itself generates some complexity (Schelling, 1961).

The assumption that the players' choices in the game are independent from their context in the real world has, however, been proved to be wrong in some cases (Daré, 2005); when the game is known, for example, some players introduce strategic behaviours to lead the ex-post discussions in certain desired directions. The less the players are constrained through the rules of the games, the more they will find opportunities to invite real word issues into the game. In some cases this is even the purpose of the game (D'Aquino *et al.*, 2003). These ad hoc behavioural patterns should therefore be analysed through debriefing sessions and interviews. The game provides a micro-world that can be analysed on the specific issues it has been designed for, with the double axes of validation of game-designer expectations and reactions of players according to their feelings in the game, as well as to their own experience of such collective decision processes in the real world.

Concert'eau:[5] simulating collective assemblage of plurality

We based the design of the Concert'eau game on Eco-logiques (Germe and Thévenot, 1996). Concert'eau's first step is similar to Eco-logiques': making the players aware of how people differ in the values and principles they defend, and not only in their interests. Moreover, Eco-logiques is aimed at analysing which arguments are considered to be most legitimate in collective discussion. Concert'eau is supposed to be a generic representation of a collective decision process, based on a deliberative mode or hybrid forum model, with contextual elements borrowed from the case studies and rough categories of argumentation that can be observed in the Llech and Lentilla basins. It is aimed at getting players to experiment with collective decision-making and to compromise between opposing justification-related arguments.

Presentation of the game

The game includes eight players divided up into four teams ("Do", "Ré", "Mi" and "Fa"). Each team is an inhabitant of the imaginary "four seasons' valley". There is a collective discussion table at the centre of the room, as well as inhabitants' "houses": a two-person table for each team at the four corners of the room.

Viewing participant differences through various criteria of common evaluation

The four "orders of worth" for water management described above (part 2) form four "departure cards" given to each two-player team. This card presents the logic of evaluation they will have to defend when

[5] Concert'eau means collective discussion ("*concertation*" in French) about water ("*eau*") but it also evokes music and the difficulties involved in composing a good piece.

facing events and when needing to make compromises with other teams. In this way, we lead players into a position where they have to promote an "orders of worth" for water management and to categorise issues through this rationale: green "order of worth" ("Fa" team); patrimonial or domestic "order of worth" ("Do" team); market and industrial "order of worth" ("Ré" team); and civic and fame "order of worth" ("Mi" team). Departure cards that define players' roles are an incentive to shift to the justification pragmatic regime. The aim is to make people grasp how participants differ, using various criteria of common evaluation. In so doing, Concert'eau aims to get people to shift from an a priori strategic engagement in collective discussion to a justification regime.

First step: familiarising players with their role and with that of the others

The teams' roles are only defined as those of inhabitants who have to defend the principle featured in their departure card. In the first step, we give each team 34 cards, including photos, extracts from interviews (such as those chosen to illustrate river basin goods above), or extracts from documents generated from interviews on the case. They provide information in accordance with three formats. Teams go to their "house" and choose from the 34 cards the six that in their opinion are the best "match" for their departure card. They have 20 minutes for this. In the following step they present these chosen cards to the other teams, and discuss the cards chosen collectively. The facilitator points out when cards are chosen by two or more teams, showing the possibility of an agreement but also trying to make explicit the reasons for this choice by each team. In this step, players in each team are embodying "Do", "Ré" "Mi" and "Fa", giving them some life. They are crafting their role on the basis of their own experience of dialogue settings for water management issues.

Second step: making compromises

The context is a collective meeting where players discuss how to react to events concerning some aspects of water management that they are jointly facing. They are asked to reach a compromise. Players are incited to reach a compromise through the assurance that their proposals are likely to be taken into account if they are the result of a consensus: when there is agreement, the facilitator will inform an imaginary authority so that it can be taken into account in a public decision. The game's facilitator chooses the events' progression according to a previous discussion of events so as to create problems for the players, thus encouraging them to experiment with the difficulties of combining plurality. For each "event", each team has to give its own opinion. Team members can build their opinion during a five-minute team discussion in their "house". They then come back to the collective table and have 10 minutes to discuss and reach a compromise with other teams. The game facilitator helps teams to write down the compromise. The teams can then sign it if they still agree.

Presenting events so as to provide various formats of information

Events are short texts (around 150 words) written on a sheet of paper. For example, a retired farmer plans to sell his land to an external investor who wants to build a large tourist resort. Events are written to provide elements from diverse formats of information (Thévenot, 2004), as the 34 cards did. They give information linked to the four teams' rationale through inhabitants' reactions: "Mr. Dupatelin is happy because his son may be able to find a job in the resort. Mr. Dusouci worries about water provision and environmental balance." They also provide elements that refer to proximity regimes. Each team receives the same event card, but one of them gets a slightly modified one: for the same event, this team's card introduces a reference to some personal ties. In one event for example, the nephew of a green team member pollutes the river, whereas for the three other teams it is an anonymous cattle farmer who pollutes. In another event the "Ré" team's own property is flooded, while for the others it is the property of an anonymous inhabitant that is flooded.

A few contextual descriptions

During the game session, we describe the context of the collective discussion very roughly: we give information neither on the public authority and status of the collective meeting, nor on any social or professional status of the players. Player roles are described very roughly. Players are inhabitants with a departure card but without professions or interests to defend. They are all considered equal except for their justification principle. They are inhabitants and not just disembodied principles, meaning that defending goods other than common goods is possible. But they are defined only as inhabitants, in order to limit elements that could favour strategic behaviours. Even though this strategic behaviour is one of the regimes of the theoretical framework presented in the first section, a strong orientation towards this behaviour was observed in the first test of the game. Players tried to embody themselves in a socio-professional category that they considered as the clearest archetype: farmers and fishermen. When they had adopted their representation of this archetype, they attempted to defend its interests.

Finally, limiting the information available leads the players to ask for further elements of context or to import them from their own experience, in order to act in the game. The observation of these additions provided insights into what is required for hybridation to take place in dialogue settings and what information and guarantees participants need before they commit to the collective discussion.

Third step: debriefing

After three or four events, players are asked to step back into their "own shoes" and to leave their "Do", "Ré", "Mi" or "Fa" shoes. The game facilitator's job is over; the game observer shifts to being the debriefing facilitator. The debriefing discussion deals with the participants' feelings during the game and provides a return to reality. It is organised around the following questions:
- How did you feel during the game? Did you feel at ease? What difficulties did you encounter during the game?
- During the events, who do you think you were? Where do you think you were?
- If you played the game again, would you play the same way?
- Do you think what happened in the game could happen in reality?

One or two months after the game, an individual debriefing takes place with some players, in line with observations during the game.

This game does not aim to get people to change their values or analyse such changes, as in (Kergreis, 2004), or analyse which arguments are considered most legitimate. Nor does it aim to analyse strategies in negotiation situations, but rather to allow people to experience the difficulties of combining differing but potentially equally legitimate values and other goods involved in the game. Its objective is to subject the players to the tensions encountered when plurality needs to be combined, a major challenge of democracy.

First results of the game tests

Three tests were carried out involving students and colleagues (engineers and scientists from different disciplinary backgrounds). We present the results here before drawing hypotheses in the next section, in the prospect of a game session with real stakeholders.

The difficulty of embodying values and shifts in socio-professional classes

During the test, many players underlined the difficulties they had embodying the principles, and they even expressed some inability to do so. A player from the "Mi" team said: "We are elected representatives… In fact, we are trying to translate our management principle. It's easier to reason following an archetypal figure than following a management principle." Many players felt troubled by having no information on the stake they hold or their profession: "Each of us is necessarily a representative of something while events occur." Players who defined their role as stakeholders defending an interest ("I play chess") were the most uncomfortable. The potential for each regime to occur differed among participants. Some claimed that they could not compromise with principles: "The problem is that you cannot negotiate or compromise with principles. A principle is a principle. To make compromises you need to get beyond principles. Principles involve intellectualising, and people are not that intellectual." This player doubts people's capacity to talk about principles. Concert'eau assumes the need and the capacity to talk at the three different levels: interests, values and proximity.

Players contributed to the game and to their role. They specified their roles so that they were easier to embody. Faced with events, players quickly switched from values to socio-professional classes and identified groups. They used what they thought were categories of water management: participant "Fa" was named "*radical-ecologist*" or "*environmentalist association*". In the same way, "Mi" was named "*civil service*" ("*l'administration*"), "*mayor*", "*elected representative*" or "*public authorities*". "Do" was named "*farmer*", "*traditional farmer*", "*old person*" or "*retired mayor*". "Ré" was named "*hotel-keeper*", "*intensive farmer*" ("FNSEA"), "*developer*" or "*planning officer*". The causal link between events favours the maintenance of the role embodied as a farmer.

Teams had different degrees of difficulty remaining in the justification regime. The "Do" team, which was the domestic one, easily embodied a "farmer" and scarcely acted in a justification regime. In tests one and three, some players embodied stakeholders and could not go back to the justification regime. They stuck to defending their interests and attracted the other teams to that way of engaging with the situation. Then other teams defended interests or remained silent, as "Mi" did in test three. "Fa" easily stayed within a regime of justification. The name other players gave favours this regime. But as green good is today expressed in public spaces, people experienced no difficulty with this regime (except if it was too far opposed to their own thinking). "Mi" was also forced to stay in the regime of justification because of the imaginary authority, represented by the game facilitator. "Mi" could not embody a representative of public interests.

Different ways of dealing with proximity ties

"Do" team players never noticed when they had the "personal event" (a cousin wants to sell his property for a big tourist resort). At the same time "Do" never felt uncomfortable. In the three test sessions, they had no difficulty speaking about personal ties. When facing the "Do" players, other teams' players felt obliged to use personal arguments to prove that they "knew" the river basin. One "Ré" player said: "Do played on the local ties. So we were all obliged, at one time or another, if we wanted to be listened to, to put forward our ties. I had to say I had five children." In the 3rd session, "Do" said: "We thought that the 'Mi' players were strangers to the river basin", underlining the difficulty the "Mi" team encountered moving beyond principles. A "Do" player asked other teams to prove their proximity knowledge of the river basin.

In contrast, other teams felt uncomfortable when confronted with personal arguments in the first step of the game that involved personal testimony and later involved the "personal event". Various reactions were observable. During the events, the family or personal link was not used in the same way by all the teams. The two members of a given team sometimes disagreed on the position to adopt. In the 2nd test, one player from team "Fa", who appeared to be a leader during the second event discussion, spoke little during the third event discussion involving his nephew. His partner revealed the personal link at the end of the discussion when all the teams had reached agreement. In the 3rd test, "Mi" decided to conceal personal links to a canoe renter involved in one of the events. They excluded this fact from the discussion and criticised those who took into account this category of relationship in the public space. A "Mi" player said: "When you ('Fa') defended your nephew, I felt it was contradictory to your environmentalist role. I don't know if it's right. Should your family ties come before team principles?" For him the collective is more important than the individual, individual interests and personal matters should not appear in public spaces. In another test session, faced with the nephew polluting the river, "Fa" expressed unease at the collective table. In the three tests, the nephew's polluting behaviour forced the team to shift from the justification regime, because they felt it was intolerable to maintain a strong position of environmental defence. This shift facilitated the compromise. But the two players of the Fa team disagreed. One of them stressed the necessity of using such arguments because personal links should facilitate a fair solution due to information transparency. Another one "trying to think according to an environmentalist family philosophy" told other players: "My nephew cannot be polluting, he must be an environmentalist."

Discussion and perspectives

Participants' roles in water management settings

According to preliminary experiments, Concert'eau provided answers to the following questions: How do people act in collective discussion situations? What are they expecting from other participants? These questions concern the roles of participants in water management settings. The role is a key sociological concept, which links individuals and institutions. The first meaning of role is normative: a social constraint. It is a set of normative expectations attached to a social position. But people distance themselves from their role and may have several roles. Goffman (1991) makes a careful distinction between "typical role", "the normative aspects of role", and a "particular individual's actual role performance". Role in the normative sense is to be distinguished from role performance or role enactment, which is the actual conduct of a particular individual while on duty in his or her position. Typical dimension means expectations of a role such as the set of socially agreed-upon assumptions about the behaviour of people in certain social situations. It deals with the representation of the social world and with its division into groups and classes.

Concert'eau allows the performance process to be observed and reveals preconceived opinions. The normative dimension is specified in the game's rules. "Typical roles" appear during games, such as an environmental association or a traditional farmer. Finally, players' roles are crafted during interaction. The limited descriptions of the context and of the role make players fulfil their role and embody it. Debriefing gathers information on the "types" handled and the role's embodiment. We use questions such as: "Where do you think you were?" and "Who do you think the other players were?" to link the situation and the "types" handled. The three tests showed that players were at ease with some socio-professional "types" and uncomfortable with the role-principles proposed in the departure cards. Socio-professional classes appear to be roles with associated normative constraints that are well known, or supposed to be, by the players.

The theoretical framework presented in section 1 proposes categories to describe role design from a different viewpoint: one participant may expect another to be a stakeholder, a close relative or friend or a person with different values. Some players expected others to be mainly stakeholders and found it difficult to talk about values. This is in line with the results of Kergreis (2004) who carried out experiments comparing various evaluative and descriptive pieces of knowledge of field boundaries between farming students and environment students. She analyses the conflict between "fact norms" (practices) and "injunctive norms" during periods of social change. She observes that participants in these experiments stick to the identity of their social group and that no discussion on values occurs. She compares experiments to real collective discussions and points out that there is no real collective discussion or questioning of participants' values in both cases. The use of Concert'eau provides insights into the expectations of participants and critical feedback on the framing induced by participatory settings.

Proximity knowledge and deliberative model

Concert'eau also addresses the issue of including what we name "proximity goods" in the assembling of plurality for a collective decision. A deliberative model assumes decision fairness requires publicness. A public or private categorisation would lead to proximity being treated as a black box. This in turn would lead to pushing proximity knowledge out of the public space. We prefer to consider a continuum from proximity to public stances and study the tension involved in moving from one to another. Dialogue settings may pave a two-way road from proximity to publicness. The three tests showed players had various ways of taking into account personal elements in the assembling of plurality. We assume that three factors influenced the way that people take into account personal elements: the game departure card, perceptions of the place embodied by the collective table in the game, and the players' own political experience of public debate.

It seemed easier for "Do" teams to take proximity knowledge and ties into account. We make the assumption that difficulties in integrating proximity knowledge depend on game departure cards or on "orders of worth". A domestic "order of worth" makes personal ties general through fraternity. It is a common good built to protect personal ties. This facility sometimes paves the way for criticism of "paternalism". People also find a way towards other common evaluative criteria to value proximity knowledge. Doidy (2003), for example, studies the collaboration between fishermen and ecologists, using fishermen's knowledge through "water sense" as arguments for public debate. They make general arguments from other orders of justification than the domestic, such as the green.

We argue that the context of the assembling process influences the way people take into account proximity knowledge. In the debriefing, we were able to specify how people considered the game setting (publicness of the place...). In the test sessions, several players felt uncomfortable because they had no clear representation of what this situation was supposed to be. The simulated dialogue setting was understood as a local water commission assembly or an advisory committee. This representation of the situation influenced the way that they actually allowed proximity knowledge to be voiced.

The players' differing abilities to bring proximity knowledge into public depends on various political models. A "Fa" player in one session used the community solution by claiming his nephew couldn't be polluting because it was not in his family philosophy. A "Mi" player in the same session stated that "private concerns should not be invited into the debate", which means that proximity knowledge is not suitable for public discussions. Whereas Test 1's "Fa" players or Test 2's "Mi" players spoke first in the event discussion, directly and publicly expressing their embarrassment over their relatives' action. Discussions among players of the same team about what to do with the "personal event" provide material for the study and comparison of peoples' ability to make proximity general and their political means of doing so.

Conclusion

The Concert'eau role-playing game is designed as a collective discussion model and experimental tool to gain insights into how people act in collective decision processes: it leads participants to implement justification principles in their behaviour. This tool has notably raised the issue of how participants expect and "perform" roles in collective discussions about water issues and their preconceived opinions about other participants. Three tests with Concert'eau showed that players felt uncomfortable with the roles or principles proposed in the departure cards, while they felt at ease with some "typical roles" in the water field: categories of users like farmers or fishermen. The issue of shifting to proximity appeared a tricky one in the experiments.

The specific uses of Concert'eau and its working hypotheses can be reconsidered when playing it with people who have real interests and attachments to the river basin. For the three tests with students and colleagues, Concert'eau was only an experimental setting. On the Lentilla and the Llech, Concert'eau provided an additional piece of the water collective dialogue setting, since crafting a new river basin collective institution is envisaged with publicness and participative issues.[6] Findings from the three test sessions could constitute useful input when collaborative water management settings are being designed. Tackling the role issue on the one hand and the proximity issue on the other could provide insights into the following questions: What normative role and condition can each piece of the setting propose to participants? Roles

[6] Tests with people linked to Llech and Lentilla water were carried out and resulted in a revision of initial questions presented in this chapter for further use in the field. See Richard-Ferroudji, 2008b

could be defined with reference to the three levels: principles, interests and personal ties. How could the setting favour shifts between proximity and public stances? Beuret and Doidy (2001) pave the way to answering this second question. For example, they point out the role of mediators as links (*passerelles*) to bridge the gap between proximity and public. In practice, this means, for example, employing facilitators for river basin organisations, as observed from other case studies (Richard-Ferroudji, 2008a).

References

Agence de l'eau Rhône-Méditerranée-Corse, DIREN Rhône Alpes, *SDAGE RMC, Guide technique n° 8 : eau et aménagement du territoire en RMC*, Lyon, 2003.

Allain S., Emerit A., "Projets d'aménagement hydraulique, concertation et planification participative de bassin : une approche en terme d'action publique négociée", *Natures Sciences Sociétés*, 11, 2003, pp. 255-265.

Babin D., Bertrand A. *et al.*, "Patrimonial Mediation and Management Subsidiarity: Managing Pluralism for Sustainable Forestry and Rural Development", *Pluralism and Sustainable Forestry and Rural Development Proceedings of an International Workshop*, Rome, FAO-IUFRO-CIRAD, 1999.

Barreteau O., "The Joint Use of Role-playing Games and Models Regarding Negotiation Processes: Characterization of Associations", *Journal of Artificial Societies and Social Simulations*, 6, 2003, http//jasss.soc.surrey.ac.uk/6/2/3.html.

Beuret J.E., Doidy E., "La concertation comme émergence : le cas des usagers de l'espace rural", *XIX*[e] *European Congress for Rural Sociology*, Dijon, 2001.

Boltanski L., Thévenot L., "Finding One's Way in Social Space: a Study Based on Games", *Social Science Information*, 22, 1983, pp. 631-680.

Boltanski L., Thévenot L., "The Sociology of Critical Capacity", *European Journal of Social Theory*, 2, 3, 1999, pp. 359-377.

Boltanski L., Thévenot L., *On Justification: Economies of Worth*, Princeton, Princeton University Press, 2006 (2001).

Bousquet F., Barreteau O. *et al.*, "An Environmental Modelling Approach. The Use of Multi-agent Simulations", in Blasco F., Weill A. (eds.), *Advances in Environmental and Ecological Modelling*, Amsterdam, Elsevier, 1999, pp. 113-122.

Bousquet F., Barreteau O. *et al.*, "Multi-agent Systems and Role Games: an Approach for Ecosystem Co-management", in Janssen M. (ed.), *Complexity and Ecosystem Management: the Theory and Practice of Multi-agent Approaches*, Cheltenham, Edward Elgar Publishing, 2002, pp. 248-285.

Broc N., Brunet M., Caucanas S., Desailly B., Vigneau J., *De l'eau et des hommes en terre catalane*, Perpignan, Trabucaire, 1992.

Callon M., Lascoumes P., Barthes Y., *Agir dans un monde incertain, essai sur la démocratie technique*, Paris, Seuil, 2001.

Cheyns E., "Formats d'enquêtes. Expériences de recherche autour des processus de qualification des produits alimentaires en Afrique", in Eymard-

Duvernay F., Favereau O. (eds.), *Conventions et institutions*, Paris, La Découverte, 2005, pp. 401-418.

Cockes I., "Mediation Support for Forest Land Allocation: the SIRO-MED System", *Environmental Management*, 20, 1996, pp. 41-52.

D'Aquino P., Le Page C., Bousquet F., Bah. A., "Using Self-designed Role-playing Games and a Multi-agent System to Empower a Local Decision-making Process for Land Use Management: the SelfCormas Experiment in Senegal", *Journal of Artificial Societies and Social Simulations*, 6, 2003, http://jasss.soc.surrey.ac.uk/6/3/5.html.

Daré W.,"Comportements des acteurs dans le jeu et dans la réalité : indépendance ou correspondance ? Analyse sociologique de l'utilisation de jeux de rôles en aide à la concertation pour la gestion de l'eau", *Sciences de l'environnement*, Paris, ENGREF, 2005.

Doidy E., "Faire entendre la voix des usagers dans les concertations environnementales", *Sociologie pratique*, 7, 2003, pp. 49-64.

Dziedzicki J.M., "Gestion des conflits d'aménagement de l'espace : quelle place pour les processus de médiation ?", *Aménagement de l'espace et urbanisme*, Tours, Université François Rabelais, 2001.

Etienne M. (ed.), *La modélisation d'accompagnement : une démarche participative en appui au développement durable*, Paris, QUAE, 2009.

Faure G.O., Mermet L., Touzard H., Dupont C. (eds.), La négociation – situations et problématiques, Paris, Nathan, 1998.

Ferber J., *Multi-agent Systems, Reading*, London, Addison-Wesley Longman, 1999.

Germe J.F., Thévenot L., "Le traitement local des conflits en matière d'environnement Jeu éco-Logiques", *Un jeu pédagogique sur les logiques d'argumentation dans les conflits autour de projets d'aménagement*, rapport final 2, La Défense, Institut international de Paris, 1996.

Goffman E., *Les cadres de l'expérience*, Paris, Minuit, 1991.

Kergreis S., *Régulations cognitives et sociales dans les concertations agri-environnementales : effets des contextes sociaux et des supports de discussion sur l'évolution des connaissances descriptives et évaluatives des bordures de champs agricoles*, thèse de psychologie sociale, Université de Rennes, 2004.

Kraft M.E., Johnson B.N., "Clean Water and Promise of Collaborative Decision Making: the Case of the Fox Wolf Basin in Wisconsin", in Mazmanian D.A., Kraft M.E. (eds.), *Towards Sustainable Communities: Transition and Transformation in Environmental Policy*, Massachusetts, MIT Press, 1999, pp. 113-151.

Mermet L., "Place et conduite de la négociation dans les processus complexes : l'exemple d'un conflit d'environnement", in Faure G.O., Mermet L., Touzard H., Dupont C. (eds.), *La négociation. Situations et problématiques*, Paris, Nathan, 1998, pp. 139-172.

Moody M., Thévenot L., "Comparing Models of Strategy, Interests, and the Public Good in French and American Environmental Disputes", in Lamont M., Thévenot L. (eds.), *Rethinking Comparative Cultural Sociology: Repertoires of Evaluation in France and the United States*, Cambridge, Cambridge University Press, 2000, pp. 273-305.

Moss S., Davidsson P., (eds.), *Multi-agent-based Simulations*, LNAI Series, Springer, 2001.

Mucchielli A., *Les jeux de rôles*, Presses universitaires de France, 1983.

Richard-Ferroudji A., "Vers l'institutionnalisation d'une 'gouvernance de bassin' ? Le cas du bassin versant de l'Orb", *Séminaire PCSI*, Montpellier, CIRAD, 2003.

Richard-Ferroudji A., "L'animateur de bassin versant : insuffler vie à une communauté de l'eau", *Cosmopolitiques*, 17, 10, 2008a.

Richard-Ferroudji A., *L'appropriation des dispositifs de gestion locale et participative de l'eau – Composer avec une pluralité de valeurs, d'objectifs et d'attachements*, thèse de doctorat de l'EHESS, spécialité sociologie, Paris, 2008b.

Ruf T., "Usages et droits d'eau dans les canaux de montagne. Sept siècles d'histoire à Prades (Pyrénées-Orientales)", *Histoire et sociétés rurales*, 16, 2000, pp. 11-44.

Salles D., Zelem M.C., Busca D., Gendre C., *Les stratégies des agriculteurs face aux dispositifs de gestion de l'eau*, Toulouse, CERTOP-CNRS GIS ECOBAG, 1999.

Saunders-Newton D., Scott H., "'But the Computer Said!' Credible Uses of Computational Modeling in Public Sector Decision Making", *Social Science Computer Review*, 19, 1, 2001, pp. 47-65.

Schelling T.C., "Experimental Games and Bargaining Theory", *World Politics*, 14, 1961, pp. 47-68.

Selin S., Chavez D., "Developping a Collaborative Model for Environmental Planning and Management", *Environmental Management*, 19, 2, 1995, pp. 189-195.

Thévenot L., "Pragmatic Regimes Governing the Engagement with the World", in Knorr-Cetina K., Schatzki T., Savigny Eike V. (eds.), *The Practice Turn in Contemporary Theory*, London, Routledge, 1999, pp. 56-73.

Thévenot L., "Which Road to Follow? The Moral Complexity of an 'Equipped Humanity'", in Law J., Mol A. (eds.), *Complexities in Science, Technology and Medicine*, Durham and London, Duke University Press, 2000.

Thévenot L., "Les enjeux d'une pluralité de formats d'information. Du partage au marché", in Eric D. (ed.), *Regards croisés sur la circulation du savoir*, Lille, Septentrion, 2004, pp. 333-347.

Thévenot L., Moody M., Lafaye C., "Forms of Valuing Nature: Arguments and Modes of Justification in French and American Environmental Disputes", in Lamont M., Thévenot L. (eds.), *Rethinking Comparative Cultural Sociology: Repertoires of Evaluation in France and the United States*, Cambridge University Press, 2000, pp. 229-272.

Tsuchiya S., "Simulation/Gaming as an Essential Enabler of Organizational Change", *Simulation and Gaming*, 29, 1998, pp. 400-408.

Importing Democracy

Promoting Participatory Decision Making in Russian Forest Communities[1]

Maria TYSIACHNIOUK* and Errol MEIDINGER**

**Researcher, centre for independent social research, St Petersburg, Russia,*
***Professor, State University of New York at Buffalo, USA*

In the 1990s, after Perestroika opened up the borders of the former Soviet Union, Western interests acted quickly to link Russia into global markets and institutions. An array of multinational companies promoted the restructuring of Russia's legal and economic infrastructure to facilitate their operations in the country. Large transnational environmental organisations established active subsidiaries in Russia as quickly as commercial interests did. Bringing with them Western money, values and knowledge, these organisations officially sought to become important players in Russia's political and economic spheres.

For modern approaches to forestry to be imported, management practices developed in the West needed to be adapted to Russia's unique post-Soviet context. For example, many of the social aspects of sustainable forestry, such as community participation in forestry decision-making, found little pre-existing social infrastructure in rural Russian forestry communities, making for a significant institutional challenge.

In the 2000s the official period of transition to the market economy came to an end, as did the inflow of funding for building democratic institutions. Yet the expectations of global markets, particularly European ones, continued to include a high level of community participation in forest management. To sustain their growing involvement in

1 This work was sponsored by the MacArthur Foundation in 2002-2003, and by a grant from the Academy of Finland (No. 208144) in 2004-2005 as part of the project on "Governance of Renewable Natural Resources in Northwest Russia". In 2006-2007 the project was supported by the Moscow Public Scientific Fund and in 2008-2009 by a grant from the Academy of Finland (No. 121428) as part of the project "Transnationalisation of Forest Governance".

international trade, forestry holdings in Northwestern Russia became involved in Forest Stewardship Council (FSC) certification; FSC institutional designs and rules of the game were institutionalised in a continually expanding area. Russia now has the second most FSC certified land in the world, indicating that community participation in local forest decision-making should have increased considerably as well. Based on longitudinal research conducted from 2002 to 2009, we can now assess the outcomes of Western efforts to import democracy to Russian forest settlements. While we find that the adoption of forest certification has indeed led to significant increases in community participation in some forest management areas, that process has required a complex amalgamation of western practices and assumptions with pre-existing Russian ones. This was done to a large extent by engaging pre-existing community authorities in expanded networks of participation.

The lack of pre-existing civil society infrastructure made the transplantation of FSC institutional designs difficult in Russian villages. Many of the social aspects of FSC certification, primarily community participation in forestry decision-making, are supposed to be built from the ground up. This created a major hindrance to environmental organisations, which were trying to import democratic institutions to Russian forest settlements from more environmentally and socially advanced countries and to build democratic institutions from the top down. Civil society in the villages was limited to the existence of a few social and youth activists, hunting societies, veterans' organisations and teachers. These groups did not traditionally cooperate with one another and could not mobilise as a mass in order to express their grievances to governments and businesses. Neither were they accustomed to participating in forest management processes; they had no tradition of acting as real stakeholders in the surrounding forests.

This paper shows how the World Wildlife Fund for Nature (WWF) used market forces to jump-start democratic institutions in Russian rural communities to create a basis for social, environmental and economic modernisation within Russia's forestry sector. To analyse this effort at building grassroots democracy we describe the networks, local communities and cultural understandings – "social imaginaries" – involved in instituting more democratic management practices in Russian forestry.

Actors and networks

Transnational: Forest certification has been promoted by transnational NGOs as a way of institutionalising sustainable forest management around the world. Its transmission to communities in Russia relied on agents acting at the transnational, national-regional and local levels. The dynamic that made this possible was transnational – the rapid incorporation of Russia into global timber markets following Perestroika. Multinational corporations purchased Russian subsidiaries and established logging operations in Russia. Russian holding companies were formed and became engaged in trade on European markets. Global NGOs responded by promoting forest certification and providing incentives for sustainable forest management that include stakeholder participation as an integral part. NGOs did this in two primary ways. First, they created transnational-national-local coalitions in favour of sustainable forest management and certification. Second, they used international markets to threaten the viability of Russian exports. To a remarkable degree, the transnational NGOs played differentiated roles in the process that reflected their different organisational cultures. Thus, the WWF took the primary role in building intersectoral partnerships. As will become evident in the case studies below, the WWF supported the growth of a variety of national and local NGO networks, as well as the forestry research and development necessary to buttress them.

Greenpeace and the Taiga Rescue Network, on the other hand, focused more on directly challenging environmentally harmful forest practices – particularly the destruction of old growth forests – and raising costs for those carrying them out. It did this largely by threatening Russian access to "green" European markets if Russian timber was not produced in an acceptable manner (Tysiachniouk and Reisman, 2006; Tysiachniouk, 2009). Thus, Greenpeace can be seen as the "bad cop" threatening to punish violators of sustainable forestry standards while the WWF can be seen as the "good cop", giving them a way of bringing their operations into compliance and retaining their markets. The full picture is of course more complicated, with the WWF also playing more aggressive roles at times. But nonetheless, at the general level a division of labour is apparent. The result of this coordination is that a number of Russian companies have been impelled to develop local environmental and social policies consistent with international standards – thus locally institutionalising the global processes of forest certification.

National-Regional: Both the WWF and Greenpeace have established Russian offices and supported other partners in the course of promoting forest certification in Russia. The forest products companies and their trade networks were also important national and regional actors. These economic actors tend to behave more like negotiating, mutually adjusting partners than like unitary economic actors, but variations among them have played an important role in shaping the effects of forest certification. In the process, a new organisational field[2] has been estab-

[2] By "organisational field" we mean an interconnected group of organisations that are aware of and interact with each other, ordinarily with the assumption that they are

lished. Among the most important of these organisations were local ones that formed the FSC institutional infrastructure, including a Russian national FSC initiative, with the tripartite environmental-social-economic chambers characteristic of the international FSC system, as well as several regional working groups. The creation of these initiatives fostered dialogue between businesses and NGOs at different levels, both federal and regional. An FSC national office and FSC consulting community have been formed. All of these organisations were engaged in adapting the FSC's International Principles and Criteria (FSC n.d.) to Russian conditions by defining locally applicable indicators and verifiers (Tysiachniouk, 2006). Finally, in 2008, the Russian national FSC standard was accredited, with a strong emphasis on public participation in forest management. As FSC standards contradicted Russian national legislation, NGOs tried to change Russian legislation to conform to international standards. In order to change the legislation, both business actors and NGOs were engaged in dialogue with governmental agencies about the Forest Codes of 1997 and 2007. Therefore, the FSC significantly fostered stakeholder participation in national forest policy.

Local: timber is harvested at the local level, of course, and much important social activity in forestry is necessarily organised at this level. Important local actors include the Lesnichestva – local governmental units responsible for control over forest management in the district –[3] and the various local Forest Enterprises responsible for organising harvest activities (under the close supervision of leskhozes) and delivering timber. Finally, there are a variety of other local community groups and civic initiatives – ranging from educational institutions to advocacy groups that affect and are affected by forest certification.

Communities

While our research focuses on local "communities", using this term necessarily involves significant difficulties. The main problem is that the word carries heavily romanticised and nostalgic connotations, often invoking a traditional, stable, authentic and impliedly "good" and environmentally appropriate set of relationships maintained by relatively innocent, uncalculating individuals, in face-to-face interaction with each other. We are not prepared to include these connotations in our use of

---

somewhat interdependent and are part of a common larger process. See, e.g., Dingwerth and Pattberg 2009.

3 Leskhozes survived from the Soviet era, when they combined oversight and limited operational functions. The Forest Code of 2007 transformed them into Lestnichestvo that are allowed only to oversee forest leases, and are not yet allowed to undertake forest operations.

the term. Rather, we use it in the minimalist sense to refer to "patterned interactions among people in a local geographic setting". We are thus using a form of the "field of interaction" conception of community advocated by Roland Warren in his later work to replace the "concrete collectivity" conception with which he was originally associated (1978: 417-418).

Several related points are important. First, while it may be true that people always interact locally, and that community therefore always exists (Wilkinson, 1991), there can be enormous differences in the content and quality of community life; we are trying to see whether the introduction of forest certification affects these variables. Second, it is helpful to characterise communities in terms of vertical and horizontal integration as advocated by Warren, but it is not necessarily the case that there is a linear trade-off between the two. The work of Warren and many other community sociologists sought to connect a decline in community autonomy to the rise of centralised governments and national and multinational corporations during the "great change". While these developments may indeed have decreased the self-determination of many local communities, over the longer term they have also triggered a variety of counter-reactions, particularly in the past two decades, seeking to revitalise communities. Forest certification is a case in point, as transnational market relationships have been used to try to leverage increases in the voice and power of local actors. Whether this effort might turn out to be fruitless in the long run is impossible to say, but our data indicate that it may be significant in changes in community political interaction and understandings of how communities should be engaged in forest policy-making.

Social imaginaries

The effort to develop a local "civil society" that can affect forest policy where such a thing did not exist previously raises the question of new cultural understandings. Local communities can act systematically only to the extent that their actions are guided by intelligible images of appropriate social processes. In Russia, forest certification is trying to create something new in local communities, and it must do so by creating plausible cultural models. In our view, both community and environmental sociology have been overly conservative and often reductionist in dealing with new cultural understandings. Much environmental sociology, for example, has taken the cultural understandings of environmental movements to be natural responses when they connected environmental problems with rational self-interest. Similarly, community sociology has tended to take understandings of community as a natural outgrowth of community interaction – as a more or less natural given (e.g. Wilkinson, 1991). Since we wish to understand the use of alternative interaction models to change community dynamics in a non-reductionist way, we are drawing on the emerging concept of "social imaginaries" as articulated by Castoriadis (1987), Gaonkar (2002), Taylor (2004), and others. While there is nothing stunningly original about this awkwardly named concept, the basic idea is that social groups organise themselves with images of how people should relate to each other, and that these images take on a life of their own. Social imaginaries are not generally closed and determinate, but rather are fairly open and amenable to innovation. Thus, groups with different traditions will draw on common ideas and produce similar yet distinct social practices. This is the process we are seeing in the establishment of forest certification. Major concepts, such as old growth forests and public consultation, have been brought into Russian communities. When they have been successfully implemented they appear both to have changed the social imaginaries of those communities and to have been implemented in distinctive ways, reflecting pre-existing understandings.

Methodology

A case study approach was used in this paper. Similar cases in two regions of Russia were selected and compared in order to investigate the role of NGOs and other transnational actors in building democratic institutions in Russian rural settlements. In each of the case study areas, three field expeditions were conducted (from two weeks to two months each) in 2002, 2006 and 2007. During these expeditions, semi-structured interviews and participant observation were used. The researcher attended working groups, stakeholder meetings, public hearings and other events. Field notes were maintained and used for the analysis. The data includes 47 individual interviews for the Pskov Model Forest and 68 for the Preluzie Model Forest. The interviews were held with major groups of stakeholders, representatives of forest management units, governmental agencies of different levels, NGOs, local activists and business representatives.

Case studies

The Pskov Model Forest

Context: the Strugy-Krasnie region has a population of 18,500, about half of whom live in the regional centre, Strugy-Krasnie. This settlement is in the Pskov oblast[4] and lies 68 km northeast of the city of Pskov,

[4] An oblast is an administrative unit of the Russian Federation. Oblasts are further subdivided into districts. "Oblast" is often translated as "province".

which is in far western Russia near the Estonian and Latvian borders, about 250 km south-south-west of St. Petersburg. Before Perestroika, much of the economic activity in the region consisted of work for St. Petersburg, Moscow or Riga enterprises specialising in the Soviet military-industrial complex. Since the late 1980s, however, many of these operations have disbanded or become unstable. The regional economy has declined severely and there is significant unemployment. Logging companies in the region are export-oriented and make use of the good railway transportation to Latvia and Estonia. Accordingly, until 2007-2008, when the Russian government introduced high taxes on round wood, the Strugy-Krasnie district was an important raw material provider for the international timber industry of Europe. The Pskov Model Forest consists of 46,000 hectares on the Strugy-Krasnie Leskhoz, which STF-Strug, a subsidiary of the large Swedish-Finnish company Stora Enso, was leasing.

Company Characteristics: Stora Enso subsidiaries have been conducting export-oriented logging operations in Russia since Soviet times. Stora Enso was created in 1998 as a result of the merger of the Swedish company Stora and the Finnish company Enso. Enso began preparations for a Russian joint venture in Karelia in 1988. In 1990 this enterprise (called Ladenso) was put into operation.

In 1995 Stora-Enso established STF-Strug in the Pskov district and leased land for 49 years, with the goal of meeting international sustainability criteria. In the 2000s Stora Enso had a series of logging subsidiaries located in the Pskov, Leningrad and Novgorod regions, and in the republic of Karelia, and was eager to standardise its operations to help make the subsidiaries more efficient and profitable.

For a corporation in these conditions, the Pskov model forest became an opportunity not only to adjust its business to Russia with simultaneous development and introduction of innovation, but also to try to advance the development of a new normative base in Russia and to make an effort to change Russian conditions on behalf of its business. The participation of Stora Enso in the Pskov Model Forest project was an innovative strategy of business integration into another country. Stora Enso in Russia had to solve the problems encountered in the post-socialist transition period: continual reform of state governing bodies and forest legislation, institutional turbulence, and other realities of the time. From 1995 until the project ended in 2008[5] the company made significant progress in its path towards sustainability due to both FSC certification and Stora-Enso corporate social responsibility efforts. The

[5] It was closed because of high tariffs on round wood and the economic crisis of 2008.

added value of this process was public involvement in decision-making processes.

Model Forest and Certification: Stora Enso planned that STF-Strug would use Scandinavian logging technologies in Russia and meet FSC standards of sustainability. In practice, however, these standards frequently conflicted with the Russian Forest Codes of 1997 and 2007. Failing to log a whole plot, for example, and leaving behind "downed woody debris" to promote nutrient recycling and ecological regeneration, was inconsistent with traditional standards. In 2000 the WWF came to the region and partnered with the company. Together they were able to alter local views of acceptable forest practices. In essence, the WWF and Stora Enso, two monumental Western organisations, descended on a small, ordinary Russian locality and modified the commercial environment to comply with FSC standards. The WWF created a plan of action for the company based on scientific research and coordinated each move with government officials and civil society groups. STF-Strug carried out its logging according to the plan.

The Pskov Model Forest Project lasted from 2000 until 2008 with an annual budget of around one million dollars. Stora Enso contributed 20%; WWF Germany contributed another 20%, and the remaining 60% came from the Swedish International Development Cooperation Agency (SIDA). In summer 2003 the Pskov Model Forest received an FSC certificate following an audit by the US-based certification body Smartwood. Thereafter the Pskov Model Forest became an educational model of sustainable forestry.

Public participation in decision-making: Russia has no history or traditional mechanisms of public involvement in resource management, and the people have no past experience with it. From the beginning, STF-Strug experienced many conflicts in Russian localities. In general, people living near the leased land were suspicious of the foreign company, which they felt was coming to cut and send their forests abroad. In working with the community, it became the WWF's job to soothe public opposition to forestry as such by illustrating the difference between conventional Russian forestry and FSC sustainable forestry. In effect, through an extensive PR campaign, the WWF argued that by switching to the new, imported way of doing things, Russia's economy, environment, and society would benefit. The WWF used television programmes and newspaper publications, and organised seminars and workshops.

In 2000 the WWF launched a campaign to network with all stakeholders in the forest and educate them about sustainable forestry. The WWF held seminars and workshops, distributed written information about the FSC, and organised a few trips to Sweden so that government officials could study logging sites similar to those that the WWF and Stora Enso wished to set up. The Model Forest's demonstration plots became a key instrument with which to educate forest stakeholders.

The WWF established a small grants programme to pay for any creative project that pertained to the Pskov Model Forest. The WWF used the local intelligentsia as a mechanism for linking with the rest of the population. The small grants programme was focused on scientists, teachers, educators, a museum curator and librarians. These people were often community leaders who helped shape the views and practices of the rest of the community. For this reason, a social expert working with the WWF called such citizens a "golden fund", which would "help to form public opinion".[6] Teachers and educators especially helped spread knowledge and ideas, and ultimately shaped the social imaginaries of other residents and succeeding generations. The WWF brought its Model Forest, its money and its panda logo into the classroom by funding teachers' environmental education initiatives through the project's small grants programme. This included such programmes as recycling, nature calendars, computer education and a Children's Club of Friends of the WWF. The WWF contributed to the adaptation of a Swedish textbook on forestry *Principles of Sustainable Forest Management* to Russian conditions. In 2007, after its approval for use in secondary schools, the book was published and disseminated throughout northwestern Russia. In 2008 Stora Enso printed additional copies and disseminated them in schools situated close to their leases in different regions.

One of the WWF's main strategies with the small grants programme was to take activities that already existed and enhance their quality, while steering them towards environmental awareness and support of the Model Forest. During the project's lifetime, 32 small grants were financed. They were an effective tool to involve the local population in the project and a means of disseminating information about it. Grants funded ecological summer camps and environmental clubs, and even turned a traditional community holiday that involved saying "goodbye" to winter into an "environmental goodbye". One interesting advertising strategy saw the WWF sponsor a local school's soccer team. The team was called Panda, and the uniforms carried the WWF panda logo as well as the label of the Pskov Model Forest. The WWF further impressed the local population by bringing a famous football team, Zeneet, from St. Petersburg to play with the Panda team. Many people expressed excitement about this game, which also had a theme and symbol of nature. In short, the WWF used the project's extensive funds to establish the panda

6 Interview with the museum curator, 2002.

logo as a lasting visual fixture and the phrase "sustainable forestry" as a lasting linguistic fixture in the social imaginary of the Strugy-Krasnie community.

The WWF also made efforts to reshape public understanding of its role in forestry policy-making. FSC criteria demand that the local community have a voice in forestry decisions. By raising public interest in the Model Forest, the WWF laid the groundwork for official public participation. The Model Forest created a Forest Club that sought to bring all forest stakeholders together in a productive dialogue. The Forest Club met regularly throughout the project. Attendees included representatives of STF-Strug, forest management unit representatives, administrators, forest scientists, WWF staff and all interested local citizens. This arrangement, however, served more as an exchange of information between the project implementers and WWF grantees rather than public participation in decision-making.

Inclusion of the public in decision-making about forest management was a necessary measure, on the one hand, because this was one of the requirements for certification; and on the other hand, because all experts and visitors coming to the model forest from abroad were interested in questions of public participation in forestry-related decision-making: "In the West it is a favourite subject. They come and immediately inquire whether our public is involved in the decision-making process."[7] Involving the public, however, faced many barriers. For instance, the project tried to create a real, widespread interest in managing the forests; people often only became involved after their interests were hurt. Public participation, as defined under international certification norms, should be preemptive of conflict. This was hard to achieve. For example, the project tried to consider hunting interests where STF-Strug logged. They made an effort to involve hunters in the development of logging plans; still, they received little input. Hunters only raised their voices after logging plans were published and their hunting places were threatened.[8] The project implementers were themselves suspicious of the issue of public participation: "Maybe it is important to involve the public in Western countries, but here we have a different mentality."[9]

Public hearings were held to discuss the forest management plan during the state forest inventory process in 2002. Organisers used the experience of one of the World Bank's projects, asking participants to choose one of eight scenarios of landscape-environmental planning for the model forest. The discussion during the hearings resulted in a plan

[7] Interview with participant of the Project, 2002.

[8] Interview with social expert in forest certification, 2002.

[9] Interview with one of the PMF staff, 2002.

that represented a compromise between economic components, on the one hand, and environmental and social ones, on the other. At the same time, a more environmentally-oriented scenario, which provided for preservation of wood grouse mating areas, was accepted.[10] Such a model of hearings is hardly applicable to other regions, since existing rigid federal regulation in forest management has limited the range of possible scenarios.[11]

In sum, the WWF sought to use existing social networks and understandings to reshape environmental and public participation practices. The WWF understood the importance of linking new ideas and symbols to existing ones in reshaping community practices. This case demonstrates the importance of NGO sophistication and legwork for the success of Western commercial interests in obtaining Russia's natural resources. We cannot say, however, what the long-term effects of this initiative will be, as the financial support for the project ended and STF-Strug sold its interests to another company in 2010. It is thus very difficult to predict whether the local community will continue to expand its role as an active participant in local forest decision-making, although we think it likely that there will be some carryover.

The Preluzie Model Forest

Local context: the Komi Republic consists of 416,800 square kilometres just west of the northern Ural Mountains, approximately 900 km northwest of St. Petersburg. In villages throughout Komi, the economies are slow and many forest communities are dependent on forestry. Since 1917 the forestry sector has been the primary source of income, employing one third of the Republic's working population (Karakchiev, 2000).

Throughout the 20th century, inadequate reforestation practices have negatively affected both local villagers and the profitability of industrial harvesting. In the 1990s, 200,000 hectares of Komi's forests were clear-cut, while leskhozes and forest producers planted trees on 20-23,000 hectares – roughly 10% of deforested land. Between 1990 and 1994 Komi lost many of its traditional forest markets in central and southern Russia, Moldova and Ukraine, and production decreased catastrophically. Reforestation also fell to a fifth of its former level (Karakchiev, 2002).

The Preluzie Model Forest consists of 800,000 hectares in the Preluzie forest management unit territory in southern Komi. Within this

---

10 Interview with research director of the Project, March 2008.

11 Interview with research director of the Project, St. Petersburg, March 2008.

territory are permanent settlements, various industries and logging companies. The regional centre is Obiatchevo.

The project was sponsored by the Swiss Agency for Development and Cooperation with around $1.5 million per year. It lasted from 1996 to 2006. The Model Forest was actively built from 1999 to 2006. From 1999 to 2002 the project was implemented by the WWF. After 2002 it was implemented by Silver Taiga, a local NGO made up of the staff of the original WWF Komi office, thus providing continuity in management.

The Preluzie Model Forest is located in a region built on forestry, but not on exports. The Komi Republic is much farther east than Pskov Oblast, resulting in sharp differences between the two model forests. Pskov is close to Russia's European border, and so attracts export-oriented subsidiaries of multinational corporations. Preluzie is much farther away, with very limited transportation systems, and therefore offers limited near-term export potential. Roundwood (logs) cannot be economically exported, although high quality processed wood products can.

Industry Characteristics: There were several short-term forest leasers, varying in number from 12 to 17 working in the territory of the Preluzie Model Forest. These companies also varied in size and interest in trading in international markets. Forest certification was nonetheless successfully implemented because the certificate holder was a Leskhoz and was funded by grants. Mondi Business Paper, a key purchaser of wood pulp in Komi, significantly stimulated certification in Komi when it purchased the Siktivkar Pulp and Paper Mill and demanded that all its pulp suppliers be certified after 2009. A small amount of Komi's pulp wood goes to Kotlass Pulp and Paper Mill in Arhangelsk oblast, which is relatively far and involves high transportation costs. Sawed wood from Komi goes to both Russian and European markets. European markets provide an incentive for companies to get certified.

Certification: The Preluzie Model Forest obtained FSC certification through Smartwood in March 2003. During the certification process, Silver Taiga's main partner was the governmental forest management unit. The aim of the project was to certify not just the leased land of one company, as in the Pskov Model Forest with STF-Strug, but rather to certify the forest management of the entire territory. Regardless of the economic ramifications, the FSC system has gained much legitimacy with the government, which perceives the project as an important contributor to achieving sustainable forest management.

The main objective of the Preluzie Model Forest in obtaining certification was to improve the economic, social and ecological conditions of the Preluzie region by introducing sustainable forestry into forest management and social relations. One of its main goals was to establish new decision-making mechanisms for balancing the interests of the various stakeholders, including industry, government and local interests. As in the Pskov case, WWF Komi worked to establish networks with each of the interested parties and tried to engage them in an intersectoral dialogue about forest management. Thus the certification process was part of a larger sustainable institution-building effort.

Silver Taiga communicated with logging firms leasing territory in Preluzie and tried to interest them in certification. Silver Taiga's partnership with the industry intensified in 2003, when they started to help Mondi Business Paper to prepare their subsidiaries with leased territories and Forest Management Units with Mondi suppliers in order to become certified. The certification of forest management units also facilitated chain of custody certification of small logging firms in the region.

Stakeholder Participation: Silver Taiga linked with the local public through the use of educational institutions, its own educational programmes, media and discussion groups. It tried to involve the local public by promoting environmental education, self-governance structures, involvement in the Model Forest project and decision-making. Government agencies on the regional and oblast level, primarily the Ministry of Natural Resources, supported the project from its start and worked with the Model Forest in a working group.

The strategies and orientation set by this working group were then implemented on the regional level by the coordinating council of the Preluzie Model Forest. This council consisted of Silver Taiga employees who coordinated the activities of the Model Forest. This council was broken up into eight thematic groups, each with a specific focus. The innovation group worked closely with the forest management unit, hosted and organised the work of all experts on the project, implemented demonstration forest plots and put all Model Forest innovations into practice. The ecology group focused on virgin forests and biodiversity. The economy group dealt with economic questions and improving the effectiveness of forest use. The education group organised courses and training programmes. The forestry group facilitated the work being done by researchers from scientific institutions on improving forest management. The public outreach group organised discussions and tried to interest local populations in the project. The geographical information systems (GIS) group worked on producing a database and maps of the territory. The information group published bulletins and worked with journalists and the media. Representatives of the Preluzie Model Forest were part of the Regional FSC working group, preparing FSC standards specific to the Komi Republic. This group worked closely with the Russian national FSC initiative.

Important actors involved in the Model Forest include the regional administration, representatives of local groups interested in self-governance and research institutions. The Preluzie Model Forest maintained an intensive engagement with the local community. It had an environmental information centre in the library in the Obiatchevo settlement, which distributed information to all libraries in the settlements of Preluzie. The newspaper *Banner of Labor* published a special page with news related to the Model Forest. Educational seminars related to the Model Forest took place regularly, as did public hearings on forestry-related issues, when leasers were competing for land. The Model Forest also helped the local community with new technology and support, including new computers and fax machines for the libraries and new furniture, buses and equipment for the school. Also, by encouraging companies to meet the preconditions of FSC certification, the Model Forest helped to improve working conditions for some of the region's population employed in the forestry sector. One company, for example, built two new dining rooms serving hot food and constructed a small hut to shelter loggers in the forest.[12]

The project encountered similar barriers with the public to those met by the Pskov Model Forest, including a widespread suspicion of forestry in general. WWF/Silver Taiga overcame this perception by preaching the Western gospel of sustainable forestry, especially its promotion of social sustainability to better the public's lot. They circulated information through libraries and schools, created discussion clubs and used media such as television programmes, newspaper articles and art shows dedicated to loving and preserving nature. A similar small grants programme was established, funding local civic initiatives.

In order to involve the public in forestry, WWF/Silver Taiga created a club similar to Pskov's Forest Club called Shuvge Parma ("the sound of wind through the taiga forest" in the Komi language). Club meetings engaged various members of the local public, leskhoz workers, scientists and government officials in discussions about forests and their uses. One difference between this and Pskov's Forest Club was the larger size of Preluzie leskhoz and the fact that it contained dispersed villages, all of which were involved in Shuvge Parma. For this reason, the club was

12 Participant observation, meeting at Preluzye Leskhoz, 2002.

mobile and travelled to different villages throughout the region, holding meetings and promoting public participation.

An example of successful public participation stimulated by this club involved old growth forests. Here, WWF/Silver Taiga was able to mobilise members of the population to protect a pristine area that had already been rented for logging by the large company LuzaLes. While WWF/Silver Taiga first had to explain the concept of old-growth forest, it was easily accepted by much of Komi's native rural population, which is generally against industrial harvesting of any kind. WWF/Silver Taiga was also able to persuade other community groups, often by starting with the intelligentsia, as it had in Pskov. Although LuzaLes had already begun building an access road to log the plot, it gave up on most of the plot in the end, accepting a compromise that allowed it to log four small sections.[13]

One development illustrates the different possible uses of public participation. Silver Taiga, with the help of local citizens, identified places where people collect berries and mushrooms and promoted special logging regimes in these territories. They published recommendations approved by the government throughout Komi.[14]

The Club existed only until 2003, as did the small grants programme for community support. However, some community initiatives continued. A Forest Council, involving local leaders and former grantees of the project, was formed under the forest management unit. It continued to link up the general public with other stakeholders in forest management.[15]

When the Swiss Agency for Development and Cooperation stopped financing the Model Forest project in 2006, Mondi Business Paper financed Silver Taiga so that it could continue to facilitate interaction with stakeholders. Several other grants also supported Silver Taiga in their work with local communities in Komi.[16]

Overall, the certification process in Preluzie Leskhoz has led to increased public involvement in forest decision-making, and may have laid the foundations for a general expansion of community participation in policy-making. Citizens and groups have been linked into discussion and action networks that were relatively new to them, and they have enjoyed some success not only in influencing outcomes, but also in reconceptualising forests and forestry. It seems likely that they were also

13 It should be noted that while FSC standards seek to protect "high conservation value forests," they do not necessarily ban harvesting of old-growth timber.

14 Interview with the representative of the local community, October 2006.

15 Interview with a member of the Forest Council, November 2008.

16 Interview with Silver Taiga staff, May 2009.

reconceptualising economic relationships and their potential roles in them. Thus, we can postulate that certification may have played a significant role in triggering a reconfiguration of the community's social imaginaries, although describing the specifics of this reconfiguration must await further research.

Conclusion

Forest certification was promoted by specific actors with specific goals. The actors that promoted the FSC are located predominantly outside Russia, but in their efforts they also interacted with and reshaped networks, organisational capacities and social imaginaries inside Russia, even in small forest settlements in resource peripheries. Both case studies show the growth of significant new networks across local, regional and national borders, as well as across traditionally distinct social sectors. Organised around both market relationships and transnational NGOs, these networks have played an important role in defining acceptable policy and reshaping community relations in Russia.

Transnational NGO networks were essential to promoting local public participation in the cases described in this paper. WWF and Silver Taiga made the social connections and mobilised the resources necessary to propagate effective public participation in local communities. Government and business involvement were also important in some ways, but not essential to changing community practices. Government involvement was essential to getting the Preluzie forest management unit certified, but the community impact was orchestrated by Silver Taiga.

We need to consider the possibility that there is an "actor" not directly present in the communities, who, roughly put, is the imagined European consumer. In the westernmost case studies the actors' understanding of "the European market" was very important in shaping their decisions. Often, this understanding was indirect at best, with rumours and stereotypes playing as large a role as actual market actors in some cases. But in every case, this somewhat shadowy actor was called upon as an ally by some. Greenpeace, in particular, invoked it by threatening to make European consumers hostile to certain timber operations if they did not adopt more ecological practices and achieve FSC certification.

The WWF and FSC strategy of building stakeholder groups, while aimed primarily at promoting environmental protection, seems to have had broader social effects. The intersectoral dialogue and stakeholder involvement promoted by the WWF in the Model Forest cases seems to have laid the foundations for democratic institutions – particularly institutionalised stakeholder dialogue – which did not previously exist in

Russia.[17] As noted above, the WWF has acted largely through existing Russian epistemic communities: scientists on the national level, and librarians, school teachers and the like on the local level. But while using and strengthening these communities, it has also transformed them and linked them to others in new relationships. Particularly interesting has been the linkage to business interests.

Finally, perhaps the most intriguing questions posed by our research concern the longer-term effects on community relationships and decision-making processes resulting from the introduction of public participation and intersectoral dialogue into local forestry. FSC certification in Russia has grown steadily, even during the crisis of 2008-2009. It continues to provide the infrastructure for stakeholder dialogue and public participation. At the same time, funding for the WWF's interventions and Model Forest building expired in 2006-2008. We can immediately observe a decrease in the community initiatives that were previously supported by small grants. Resources have been important in the two model forests, and their withdrawal has certainly affected community dynamics. On the other hand, social institutions have a tendency to persist, and are rarely simple functions of money. Here it is interesting to note that the certification process had social effects, partially independent of which actors participated (although a strong NGO role appears to have been essential for planting the seeds of local democracy), and somewhat independent of the level of external funding. Thus it seems plausible that the certification process itself, when properly implemented to include community participation, has effects on local social institutions.

An important question for the next stage of our research is to determine whether a properly participatory certification process has long-term effects on community relationships and patterns of interaction, and if so, to work out how. The primary appeal of the concept of social imaginaries is that it focuses on the ways in which the images people use to make sense of social practices in fact enable and help to institutionalise those practices. In each of the case studies, the Model Forest

17 It is important to clarify that public participation per se is not new. People in socialist Russia were politically active, and enthusiastic about building a bright communist future. In the villages they were much less sceptical then in the cities. Villagers regularly held big collective meetings and participated in many collective institutions, including collective farms and lespromchozes – basically collective forest enterprises. The central change with forest certification is the acceptance of conflicting interests and stakeholder processes to deal with them. In the Soviet system there was only one stakeholder – the people moving together toward a bright future. There was much participation in building this future and many decisions, but no interest groups. It was presumed that businesses, citizens and government all had the same interests and were working toward the same goals.

development incorporating forest certification has helped to introduce key elements of a western social imaginary involving the market economy, public participation and stakeholder dialogue. It has also put that imaginary into practice by engaging key community members in participatory processes and social dialogue in relation to new forms of economic transactions.

It is essential to remember here that western social imaginaries are being combined with Russian ones to constitute images and expectations unique to the local context. One of the most fascinating parts of our research is that the most successful implementation of the western social imaginaries seems to have been built on socialist ones. The successful promoters of forest certification did not seek to build new social institutions from scratch. Rather, they drew upon and sought to renovate social imaginaries that were built during socialism – thereby creating recognisable but still unique new configurations. We hypothesise that these social imaginaries are likely to be more persistent as a result of this double foundation of intertwined images. If so, then the new assumptions about the propriety of community participation, intersectoral dialogue and public deliberation are likely to become well institutionalised; we can anticipate an important reshaping of forestry community life.

References

Castoriadis C. (Kathleen Blamey, trans), *The Imaginary Institution of Society*, Cambridge, MIT Press, 1987.

Dingwerth K., Pattberg P., "World Politics and Organizational Fields: the Case of Transnational Sustainability Governance", *European Journal of International Relation*, 15, 4, 2009, pp. 707-743.

FSC (n.d.), *FSC Principles and Criteria of Forest Stewardship*, http://www.fsc.org/fsc/how_fsc_works/policy_standards/princ_criteria, last visited 31 July 2004.

Gaonkar, Dilip Parameshwar, "Toward New Imaginaries: an Introduction", *Public Culture*, 14, 1, 2002, pp. 1-19.

Karakchiev, A., "First Vice-chair of the Head of theRepublic of Komi, Siktifkar", *Proceedings of the Conference on Forest Certification and Sustainable Forest Management in the North of European Russia*, 2-4 February 2000, pp. 64-74.

Meidinger E., Elliott C., Oesten G., "The Fundamentals of Forest Certification", in Meidinger E., Elliott C., Oesten G. (eds.), *Social and Political Dimensions of Forest Certification. Remagen-Oberwinter*, Forstbuch, 2003.

Petrov, A.P., "Topical Goals in Reforming the Russian Forest Administration and Management System", *Institutional Changes in Forest Management in Countries with Transition Economies: Problems and Solutions*, workshop proceedings from a conference held in Moscow, Russia, 25 February 2003, http://www.profor.info/pdf/InstitutionalChangeWorkshopProceedings.pdf .

Platteau J.P., Anita A., "Participatory Development in the Presence of Endogenous Community Imperfections", *The Journal of Development Studies*, 39, 2, 2002, pp. 104-136.

Schofield B., "Partners in Power: Governing the Self-sustaining Community", *Sociology*, 36, 3, 2002, pp. 663-683.

Taylor C., *Modern Social Imaginaries*, Durham, NC, Duke University Press, 2004.

Tysiachniouk M., "Forest Certification in Russia", in Cashore B., Gale F., Meidinger E., Newsom D. (eds.), *Confronting Sustainability: Forest Certification in Developing and Transitioning Countries*, New Haven, Yale School of Forestry and Environmental Studies, 2006, pp. 261-295.

Tysiachniouk M., "Adaptation of the Business of Transnational Corporation to Russian Business Environment on the Example of Stora Enso", *Economic Sociology*, 9, 4, 2008, pp. 56-72.

Tysiachniouk M., "Conflict as a Form of Governance: the Market Campaign to Save the Karelian Forests", in Nystén-Haarala, S. (ed.), *The Changing Governance of Renewable Natural Resources in Northwest Russia*, London, Ashgate, 2009, pp. 169-196.

Tysiachniouk M., Reisman J., "Co-managing the Taiga: Russian Forests and the Challenge of International Environmentalism", in Lehtinen A., Sæther B., Donner-Amnell J. (eds.), *Politics of Forests: Northern Forest Industry Regimes in the Age of Globalization*, Ashgate, Aldershot, 2004, pp. 157-175.

Warren R.L., *The Community in America*, 3rd ed., Chicago, Rand McNally, 1978.

Wilkinson K.P., *The Community in Rural America*, Westport, CT, Greenwood Press, 1991.

Worobec C.D., *Peasant Russia: Family and Community in the Post-Emancipation Period*, Princeton, NJ, Princeton University Press, 1991.

WWF, "Virgin Forests", *WWF Bulletin*, 1, March 2001, p. 2.

Empowerment among Family Farmers in Southern Brazil

The Social Construction of Durability as a Model for Agriculture, Rural Areas and Society

Angela DUARTE DAMASCENO FERREIRA*, Alfio BRANDENBURG*, Almir Sandro RODRIGUES**, Catherine DUMORA**, Eduardo BRITO SANTOS**, Gustavo PINHEIRO** and Osvaldo HELLER DA SILVA*

**Professors, **PhD Students, Federal University of Paraná, Brazil*

The Brazilian peasantry has throughout its history demonstrated its dissatisfaction with the unequal rules of land distribution and means of production that have prevailed in the rural world throughout the country. The working life of most peasants and small-scale family farmers can be seen as an obstinate struggle to maintain their way of life. Whether through violent movements or migratory strategies, the occupation of frontier regions or the silent negation of the hegemonic codes that have attempted to keep them down, their struggles have combined the need for survival and the desire to carry on the tradition of family farming, an opportunity that has always been sought in the name of the autonomy – however limited – it represents.

Notwithstanding the large numbers of people who have abandoned the rural space, a certain portion has persisted as family farmers and continues to struggle to "protagonise" its history, as well as that of the nation. They have come as a surprise to those who could not imagine – in these times of supposed planetary urbanisation – that such strong and contestatory voices might be heard from rural quarters, nor that such innovative strategies for social reproduction and construction of life projects might be used.

Although family farmers, the most salient actors in Brazilian rural areas, are a minority in terms of the country's demographic and occupational composition, their social, economic and cultural importance is much greater, as the numerous studies on them have shown. It is particularly in the way they have built strategies geared towards increasing their power to control their lives and actions – creating alternatives to what has been imposed on them – that family farmers have been able to affirm their singularity and relevance on the national stage. Thanks to these strategies, groups of family farmers have been able to contribute to the revitalisation of rural spaces and the re-building of rurality in Brazil.

The cases analysed in this paper are illustrative of these processes. They are based on research carried out in the state of Paraná in southern Brazil and reveal some of the ways in which family farmers have acted as protagonists: the formation of the **Rede Ecovia** *(Ecolife network)* of ecological agriculture; the creation of strategies that take an opposing perspective from the dominant positions on environmental conservation in the Environmental Protection Area of Guaraqueçaba; and the creation of the FETRAF, a new union of family farmers.

Farmers' empowerment and the reconstruction of the rural world

The theoretical macro-approaches that have guided these studies and their methods vary, since they are the fruit not of one study but of three, each carried out separately. They nonetheless belong to the same theoretical perspective on the rural world in general, and more specifically, on the rural world in Brazil. This perspective conceives the rural world as an integral part of development that "quite distinct from the ignorance of an old rural world destined to fade away and dissolve into the urban, is premised on the existence of two parallel dynamics, one rural and one urban, which are complementary, in a process of mutually configuring and deconstructing". (Jollivet, 1997: 10). It is therefore a perspective in which "rural is recognised as a category with material and symbolic characteristics possessing a singularity and a specific dynamic, even if integrally articulated with the 'urban world' within a concrete territory, or immersed in processes, networks and symbols that belong more generally to the realm of the urban. These are not rural societies of a totalising nature. Neither are we dealing with the autonomy of the rural in relation to the urban: our analytic model proposes interdependence, communication and complementarity". (Ferreira, 2002). In the Brazilian case, this perspective privileges research for the analysis of the different forms of rural world that co-exist there, but provides evidence that agriculture, and especially family farming, is the major defining factor of the rural environment in Brazil. Paraphrasing Hughes Lamarche (2000) "we can say that the Brazilian 'rural' is defined by agriculture, but not by agriculture alone". (Ferreira, 2002).

This is the perspective on the rural world that enables us to identify the reconstruction of rurality that is currently underway in Brazil.

The analytic schemes that guide the three pieces of research presented here,[1] although distinct, share a common recognition that the limits of human agency are set by the forces that have already structured social relations of class, race and gender. But, at the same time, they consider that outside those major historical transformations such limitations do not prevent local, community and daily life from being reconfigured or reconstructed as a result of the new modes of interaction collectively developed by those who have been marginalised by the hegemonic powers of capitalism and modernity. These strategies – which are the constructions of social agents and actors and their struggles – operate somewhere between reproduction and change. What distinguishes the analytic schema implemented here is their conception of the *forms* and *potential* of transformative action.

Furthermore, the three studies share a common view of the production of scientific knowledge and the relationship between science and society. They share a methodological approach that privileges participatory methods and techniques, as well as the goal of democratising access to research results. We provide several examples of this below.

In our research on farmers in an Environmental Protection Area, we found that conflicts between the former and the agents responsible for the execution of environmental policies had led to such a climate of tension in these rural communities that there was a generalised distrust of anyone from the outside, which led to silence around the issues that the researchers were interested in. In order to break through that silence, regaining the farmers' confidence was fundamental. Conventional techniques of field research, such as participant observation, questionnaires and partially open-ended interviews, did not prove to be effective. Under these circumstances, a closer and more integrated relationship with the farmers was cultivated, a special type of experience which was called "active participation", since it involved direct participation in their daily work-related activities, such as helping in the preparation of *mandioca* flour and homemade sweets, in the rice harvest and other tasks, and thus winning their trust. The words of a woman who was a member of the local leadership, in reference to an anthropologist who was part of the research team, testify to the effectiveness of the method: "That woman is different, she doesn't just shower us with a bunch of questions, she takes part in our work to understand what we do." The barriers came down and a fruitful dialogue was established, which also

1 The first research was guided by Alain Touraine's most recent work (from the late nineties). The other two are based on the work of Pierre Bourdieu.

made it possible to share research data and engage in discussion with the farmers when the field research was over.

Although it used more conventional methods of sociological research, the project on the FETRAF-Sul was developed on the basis of the researcher's prior immersion in the world of rural unionism. As a member of the advisory board of the labour federation the CUT – Central Única dos Trabalhadores – his involvement with union leaders and farmers worked in favour of sociological perception and, at the same time, allowed for a relationship of trust that made the research viable. Efforts were also made to discuss research results with members of the FETRAF-Sul, both during and after the research process.

The methods used in the study of the Rede Ecovida de Agroecologia were similar. Through participation in meetings with the farmers, researchers engaged in the processes of knowledge production that they were interested in and at the same time provided help with building an agro-ecology movement, fostering knowledge of the role that such a movement should play in society at large. This was facilitated by the production at the end of the meeting of documents that summarised the content of discussions, farmers' doubts and proposals, and the formal knowledge of the researchers and technicians who were participating in building the movement. The relationship between formal and informal knowledge was meant to establish (or improve) communication between two cultural worlds: that of the specialists and that of the farmers. It created conditions for the development of a relationship based on mutual trust and facilitated interaction and comprehension of social reality. From this perspective, those who are objects of research also become subjects of knowledge, insofar as they get involved and aid the researchers in their work.

Based on these approximations (a shared perspective on the rural world – conceptions that are in agreement on the relationship between the individual and society and on the production of knowledge and its role as it engages with social demands), we consider that certain *a posteriori* reflections on the cases studied can be carried out. Looking at the type of social protagonism that they analyse – which will be discussed in the sections below – there is one notion that appears useful to apply to all three experiences: the notion of *empowerment*.

The concept/strategy/technique of *empowerment* has been widely used in English speaking countries, with specific connotations.[2] It is

2 The term's origins go back to the mid XVII century in England, when the term was used with the legal meaning of "investing authority, authorising". Since then, it has come to be used in various other domains, to mean "to enable or allow". It was much used and disseminated by different social and civil rights movements in the United States, referring to the ability to struggle for access to power in the form of equal

generally accepted as including the mechanisms through which individuals, groups, social categories and collectivities gain access to the power to intervene in social institutions, programmes and contexts that are related to their lives and their history. In other usages, the term refers to the strategies or techniques for the realisation of this process of obtaining or building power.

In a more general sense, it is a notion that is implemented under several different usages, which are not always made explicit. The notion is used in a limited way in Brazil today, especially within the social sciences; it seems to enjoy wider use in research on health, education, gender equality and business management than in the academic literature in social sciences in general.[3]

The widespread use of the notion of *empowerment* in the 1980s was accompanied by its appropriation in a wide variety of forums and contexts. It is a polysemic notion that has served a variety of purposes. On the one hand, it has provided legitimacy to government initiatives that permit public powers to abdicate some of their responsibilities: passing the buck ("power") to local communities, who must then find their own ways to solve problems. When used as a method or strategy, it has provided a means to decentralise public and private administration without a real and effective distribution of power, while encouraging the population to commit to its success, or assume responsibility for their own disadvantages.

Within this latter perspective, *empowerment* is used in an individual or psychological sense. (Carvalho, 2004: 6). As analysed by Herrick (1995), this perspective suggests that all individuals should develop their own abilities to deal with their needs and should be "empowered" (that is, gain the necessary power) to do so. It is evident that such power is limited by the fact that our society is fundamentally unequal in the distribution of opportunities to develop these "abilities", and insofar as it permits the dominance of some over others; such structural factors work

rights. Also in the US and Canada, its most recent usage emerged: the term *empowerment*, during the decade of the 1970s, as a process, approach and method, that also contained the notion of the results to be obtained. In the 1980s the approach came to exercise great influence on studies and practices in the area of public health, education, feminist studies and organisation, social psychology, studies and organisations of the ecological movement, for the rights of sexual minorities, ethnic groups and other movements for citizenship rights. In the nineties its usage was disseminated as a method for business and public administration organisation and management. There are programmes for *empowerment* in public sector action (for an example, see the USDA site for its "Rural Community Empowerment programme").

3 For examples of the use of this term in several areas of study in Brazil, see Bentes, 2001; David, 2002; Carvalho, 2004; Becker, 2000 and Rodrigues and Santos, 2001.

as the pre-condition for whatever potential the "power" that has been handed over may actually have for creating alternative social conditions.

The second usage of the term is the one that has made headway among social movements. This notion took on two basic types of connotation. One was the approach that presupposed the ability of those who lack power to, when organised, "liberate themselves from oppressive conditions and take leadership in the processes of their own liberation, as well as to re-define and re-encounter their own identity". (Pinto, 1998). At the same time, it pointed to a series of strategies and methods of intervention that were meant to provoke reflection on the subaltern condition and on alienation, while at the same time building self-help networks and grassroots initiatives that could supply services or training to their participants.

With this idea of starting from community discussions, of initiatives beginning at the local level and of the building of spaces of action and discussion through daily practice – also leading directly to the building of more encompassing forums – the notion and methods of *empowerment* are seen as taking inspiration from non-violent movements such as the writings and practices of Ghandi and Liberation Theology (Pinto, 1998). There has been a general recognition of the importance of Paulo Freire, as a theoretical and methodological basis for the notions and methods of *empowerment*, which has led to the recent introduction of that author's work into Canadian and US university courses in the areas of social work and sociology.[4]

Many studies and social actions have been based on this perspective on *empowerment*, and have proposed the notion of *community empowerment* in order to distinguish it from other usages. In this notion, the focus is on social change at the daily life level, resulting from the creation of symmetric social relations (groups, associations, networks) through which the possibility of acquiring power through creating alternatives – "what should be" transforming "what is" – is established (Herrick, 1995). Among practices involved here we find the creation of spaces for discussion and knowledge exchange that are conducive both to increasing the value of group identity and to establishing strategies of opposition to the system or to political models that exclude and marginalise, kicking off a process of power to create alternatives. The idea that permeates this notion of *empowerment* is that power is not obtained only outside the group but also includes the inter-subjective relationships that favour intra-group symmetry.

---

4 According to Segundo Beaulieu (2005), some of Freire's works have been translated or re-edited in English, after his "rediscovery" by those interested in the notion of *empowerment*.

More recently, within the realm of anti-globalisation movements, this idea of *empowerment* has been taken up – namely, the democratisation of intra-group power as a way of fostering protagonism and struggle for change on a larger scale. In a recent book in which they see themselves as "heirs" of the Seattle anti-globalisation movements, and of new struggles against domination, Pignarre and Stengers (2005) include the notion of *empowerment* as technique – a provocation, they say, against the long-standing opposition between knowledge and technique. This is part of their philosophical approach to contemporary capitalism and the resistance to what they call "capitalist sorcery". Their aim is to demonstrate, from a pragmatic point of view, that there are different types of techniques of manipulation, those that encourage conformity, as well as those that foster reflection and reaction.

The authors arrive at this assertion after an interesting discussion on the social importance of the process of *empowerment*, as they perceive it – an exercise of pragmatism that transforms the submissiveness, conformism and malaise of the world into a force that obliges thinking/acting/feeling. In what may very well be an exaggerated analogy, perhaps we can say that they view the role that *empowerment* plays as very close to the role of socio-analysis for Bourdieu: "It brings out something that the others know, without realising that they know it" (Bourdieu, 2002: 15).[5] And this in turn creates the possibility of change.

For Pignarre and Stengers, it is not a matter of a psychological dimension linked to the process of "motivating" people who are powerless to become active and responsible. Nor is it the neo-liberal inflection given to the terms

> In the sense of undoing hard and conservative procedures that got in the way of chances for the "real actors" to benefit from the opportunities that the situation provided [...] enabling the "interested parties", once liberated from state and administrative limitations, to arrive at an agreement that permits a better defence of respective interests [...] (*ibidem*: 180).

On the contrary, this technique – or strategy, we could say – which has the attribute of being constantly reinvented in each new group that is constituted, makes it possible to establish a site of symmetrical relations where issues can be raised, and whose implicit objective is that in each concrete situation, the potential to hear, learn, generate trust and identify where action is possible should be realised, appropriating the powers of transformation, never by isolated individuals but collectively, and "honouring change as creation" (*ibidem*: 195).

5 Translated by the authors: "[...] accouche les autres de quelque chose qu'ils savent sans le savoir."

> The point is to "learn to resist submission collectively but without succumbing to the trap of collective impositiveness, attempting to create non-violent alternatives for change, at the times and places in which there is strength to do so as the weeds that spring up in the smallest of cracks are able to break through cement and concrete" (*ibidem*: 185).

This latter perspective on *empowerment* – communitarian, as a notion and "technique" for opening up for confrontation and construction of alternatives – seems to offer scope for reflection on the three cases to be briefly presented below. We have sought to verify whether the struggles and strategies used by these farmers for the construction of alternatives can effectively be considered a process of *empowerment*: to what extent their actions lead to individual or collective transformation, making them protagonists of the re-configuration of the rural milieu in which they live and participants in the elaboration of an alternative project/utopia for society, of a sustainable society.

Family farming and the agro-ecological life project

In the first study under consideration here, the aim of the research was to study family farmers who were actively constructing a collective agro-ecological project – the *Rede Ecovida* of Agroecology – that was being set up in the southern region of the country as a social and environmental alternative to conventional agriculture. The creation of the network *Rede Ecovida* began in the state of Santa Catarina, with a debate over the elaboration of an instrument for participative certification that could function as an effective alternative to certification through auditing, bearing in mind that the latter does not take the interests of small family farmers and their associations into account. In the year 2000 it came together with other institutions in the states of Paraná and Rio Grande do Sul, thus constituting a regional space for the articulation of family farmers, advisory organisations and people involved with the production, processing, commercialisation and consumption of ecological products.

In the three southern states where this network is now operating, similar to the rest of the country, the process of agricultural modernisation had unleashed a process of radical changes in the traditional format of agricultural production through the implementation of a new productive model, commonly referred to as the "productivist model for agriculture" or "conventional agriculture". The high social costs of this model, such as the pauperisation and social exclusion of small family farmers, as well as the ecological and productive degradation of rural areas, have been the object of numerous studies and are currently recognised.

The need to create systems that function as an alternative to conventional production has led to the emergence – in different countries and under different names – of alternative systems of production. Organic agriculture was the one that, in Brazil, emerged as the most popular, primarily due to the activities of the technicians who advised the different movements of family farmers. When the Ministry of Agriculture implemented its regulations for these alternative systems, the term *organic agriculture* prevailed, precisely due to its usage by technicians. It became the official name for all the diverse practices that came under the heading of alternative and ecological farming.

This process of institutionalisation of organic agriculture, beginning with the creation of the IN-007/99[6] standards, fuelled the movement for non-conventional agriculture and brought about a considerable leap in production as of the year 2000. However, at the same time, it led to adhesion by new producers who did not share the ecological principles on which the movement was initially based. Thus, the widening of the market for ecological products and the growing dissemination of organic agriculture led to organisational and agricultural practices that did not take the interests of the majority of family farmers who had originally been engaged in these alternative movements into account.

The "Rede Ecovida" Network organises these farmers and other partners around a different approach, one which in many senses echoes the ecological principles and historical demands of the movement of ecological farmers, such as the lowering of production costs and greater autonomy allied with environmental and social sustainability. At the collective level, it has developed as an alternative to the process of agricultural modernisation and its social and environmental consequences, to the extent that it proposes and implements practices that are not restricted to technical and agronomical aspects, but take into account the different spheres in the construction of a *project for society and for sustainable development*, centred around proposing formats for social relations that are less hierarchical and based on solidarity.

As an organisational and managerial strategy, the Ecovida network developed a decentralised system in which groups of farmers in a particular locale form a Regional Group that can include other institutions such as cooperatives and NGOs. Each group should include a minimum of three farming families, one of whose properties is visited monthly by the rest of the group. This dynamic of monthly group meetings ensures that farmers in the same region begin to act in a more integrated way, sharing ideas, techniques, problems, solutions, etc. In these meetings the family that is visited shows the group around its property, initiating not

6 Instrução Normativa do Ministério da Agricultura (IN-007/99).

only the process of participatory certification but the exchange of experiences among farmers. Through these methods, neighbouring farmers are continually following each other up, and in this way the properties of the entire group come to be "inspected". When the visit is specifically aimed at authorising certification of produce, the inspection is always carried out by farmers who belong to another group that is also associated with the Network. The fact that no property is ever certified by families from the same group, in addition to increasing the credibility of the certification process, allows for the exchange of experience among different groups. Thus, independently of the certification process itself, exchange of knowledge and experience and "participatory certification" are pursued.

What distinguishes this movement is the building of a collective project that incorporates multiple rationalities and that therefore does not remain restricted to the dimension of economic rationality.[7] Agroecology is an ideal that is aspired to, a utopia that functions more as a reference for what is sought than what can actually be attained at present, similar to the idea of sustainability: its importance lies in the way it provides future-based parameters for present-day practices, making the attainment of this collective ideal conceivable through the construction of the life projects of those social actors who participate in it.

The building of this project occurs through the actor's involvement as a subject who takes on identities and techniques that aid in the modification of the environment in which he/she participates and in transforming his/her life experiences into a means for obtaining freedom (Touraine, 1996: 172). Thus, the subject is constructed through critique and through breaking away from a function that has been socially ascribed. The break, which enables the actor to "become a subject", occurs through the search for achievement of his/her life project which the obligation to fulfil a socially ascribed duty did not allow. Thus the "'life project' is an ideal of independence and responsibility that is defined more by the struggle against heteronomy, imitation and ideology than by a content" (*ibidem*: 172).

Small family farmers are key actors in the network. Although family farmers are not a homogenous social category, these farmers, even in their diversity, have traits in common that spring from the singularity of the rural milieu: the unit of family production is in fact the unit where daily life unfolds; farmers, as a result of this, apply several rationales to the management of their property, not simply that of economics. This

7 In this regard, it is coherent with a critique of economic rationality as the prevailing form of rationality in modern societies, that "does away with all values and goals that from the economic point of view are irrational" (Gorz 2003, pp. 27-28).

perspective, which constitutes the logic of family farming, allows farmers to identify with the principles of agro-ecology. They find therein the support they need for the construction of a life project, taking account of several dimensions and not exclusively guided by the market logic that prevails in other systems of production.

Through the construction of life projects guided by agro-ecological principles, these farmers reconstruct bonds of solidarity and relationships with neighbours and the area in which they live and work. They do this through practices that allow for a new form of social organisation, decentralised and participatory.

Through recovery of old production practices and an alliance with new technologies of agro-ecological production, these actors reestablish forms of relating to nature that were being lost in the process of agricultural modernisation. An important aspect of this recovery is the perception of nature as an ally and not merely as a natural means of production that must be controlled.

This life project should be understood as an ideal of independence, whose essence lies in its opposition to the dominant logic than in any specific content. Thus, the agro-ecological practices of the *Rede Ecovida* producers can be understood as an *agro-ecological life project* that holds the potential for personal fulfilment in seeking to (re)gain autonomy, even if partial and relative, through control of the system of production and in the farmer's position as a social actor. In building this project, actors *gain power*, or *become empowered*, as a group, in a project of resistance and struggle that seeks the success of their individual/family projects.

Family agriculture in an Environmental Protection Area

The aim of the second study was to analyse the specific characteristics of the reproduction of family farming in areas protected by environmental legislation. The case in point here was the rural community of Batuva, in the Environmental Protection Area of Guaraqueçaba, Paraná.

The research attempted to verify whether the creation of an Environmental Protection Area in the municipality of Guaraqueçaba in 1985 had led to new forms of subordination of family farmers as a result of the restrictions that environmental protection legislation placed on their farming activities and their way of life. To put it another way, it sought to discover whether this situation was producing not just one, but rather a double blockage[8] of the development of family agriculture as a form of

8 The first blockage refers to the historical situation that Brazilian family agriculture has faced, always subordinate to large landowners, who were in turn advantaged by

social organisation (as a format for the organisation of agricultural production and as a form of social organisation in the widest sense.) Were family farmers in this situation opposing these new forms of control? And if so, how?

In order to understand how the rural world is reconstructed in this context of restrictions placed on the use of natural resources, we started from the hypothesis that the maintenance of practices considered to be traditional – part of their socio-cultural patrimony (Lamarche, 1997), built up by the first families to set up this rural community and maintained as social capital (Bourdieu, 2002) – and the incorporation of ecologically-based agricultural techniques into their mode of production, is the result of family farming's ability to adapt to diversified social, political and economic scenarios.

One of the most traditional of practices maintained by Batuva farmers, as well as by family farming in general, is planting and carrying out other activities for the farmer's own consumption. In the Environmental Protection Areas, patterns of settlement and agricultural activities that resemble the subsistence model of farming are accepted by the very laws that establish these areas, which include these farmers in the category of "traditional populations". Nonetheless, in the Environmental Protection Area of Guaraqueçaba, difficulties have emerged regarding management and inspecting authorities' recognition of the requirements of such activities. Authorisation must be obtained to clear the land once vegetation has reached a height of 20 cm, and involves a series of bureaucratic procedures. Planting is thus negatively affected, stopped from occurring at the right time, according to the local agricultural calendar; the cycle of expansion and retraction of planted areas is often obstructed, depending as it does on family demographic cycles. At the same time, traditional techniques for allowing the land to lie fallow are also obstructed. Furthermore, the imposition of severe restrictions on extractive and hunting activities, fundamental for the subsistence of farmers, makes the reproduction of the family labour force difficult.

The silent response of farmers to these impediments has been to continue to engage in these practices in a clandestine way, organising them or planning them collectively using informal channels: prevented

---

an agricultural policy that guaranteed their modernisation and ensured their reproduction by promoting the social recognition of their production model. (Wanderley, 1996). The second blockage refers to the situation created when areas governed by legislation regulating agricultural activity and prohibiting the extraction of woodland resources are not provided with alternatives for the development of agricultural activities compatible with the conservation and preservation of natural resources. This lack of options intensifies further the processes that are making small family agriculture ever more precarious (see Zanoni, Magda M. *et al.*, 2001).

from planting, they set up cultivated areas hidden in dense vegetation, which makes their work more physically difficult, involving travel to distant areas that would be harder to discover. Banned from hunting or extracting palm-heart, they continue to engage in such practices, but again, in more distant regions, which increases the physical and (moral) emotional strain involved.[9]

The negative form in which environmental legislation has been implemented in these Environmental Protection areas has promoted an objective expropriation of family farmers' practical knowledge. This knowledge was once considered legitimate but is now seen as "lay knowledge" (traditional), conducive to environmental damage. It is in opposition to "modern scientific knowledge", which is based on a knowledge of ecology, and thus legitimate and in accordance with structural demands for the conservation of nature – in this case, the preservation of the area known as the Mata Atlântica (Atlantic Forest).

This process of objective expropriation of those who are located in a position of social exclusion leads to the symbolic impoverishment of the stock ("capital") of traditional knowledge that they possess, in relation to modern, technical and legitimated knowledge.

In this regard, environmental knowledge as it is implemented in the Environmental Protection Area of Guaraqueçaba promotes further social division, separating those who possess legitimacy in terms of knowledge, and therefore practices, and those to whom such resources are denied.

Nonetheless, the rural world is being rebuilt through the actualisation of its traditional practices: by keeping up its subsistence practices, although clandestinely, at the same time as family farmers incorporate new knowledge and adjust to new contexts. This is the case for the agro-industrialisation of farm products and for the processes of reconversion to organic agriculture, as exemplified by the banana processing factory that the rural community of Batuva has set up.

This practice provides evidence of the farmer's ability to engage in alliances that permit them to better situate themselves within their social milieu. Banana sales by farmers associated with the Batuva rural community (which sells all its produce, including a large portion destined for export to Switzerland) have allowed a significant increase in income (in some cases, income from this activity has been as high as 70% of total income from the full range of economic activities the community engages in.)

9 Moral strain is a term suggested by Silva (2000) and refers to the moral stress placed on the family farmer who comes to be considered a "criminal" for engaging in traditional subsistence practices that have been banned.

Entrance into the complex export market has linked this community to a network of relations that are much wider than those that they customarily maintained with neighbouring communities and towns. It links them to a much more demanding set of economic processes in terms of the quality of the products sold and the specific type of consumer – the consumer of organic products. In addition to material gains, there are real gains in terms of "symbolic capital", such as prestige and recognition as a community of farmers that produce without harming the environment. The community thus occupies a social position in marked contrast, for example, to communities engaging in the type of agriculture that uses chemical pesticides and fertilisers.

But adhesion to these new productive practices, and the gains in terms of prestige and respect, have not led to the abandonment of practices considered to be traditional. Production for family consumption continues to exist, as well as the use of kinship relations for the exchange of services and for cooperation in particular projects where periodic help with labour is needed. Support networks continue to function, still activated when there are interests to be protected; in other words, the social capital that has been accumulated throughout these families' history has not been lost, but rather reconstructed.

This case illustrates that the "double blockage" we have referred to is not a naturally-occurring process, a historic fate only for environmentally-protected areas. Granted, we see here a specific policy of environmental protection being implemented by public powers without engaging in negotiation with the main actors present in the region: farmers and fishing communities. Yet, at the same time, it is evident that by exercising their ability to conserve important aspects of their socio-cultural patrimony, even if in doing so they must challenge environmental laws, and by incorporating innovative initiatives such as organic agriculture into their practices, local family farmers gain power, in the sense of being able to reproduce themselves materially and symbolically as farmers and as agents of environmental protection. This is informal and not necessarily a consciously-pursued form of empowerment. Through strategies dictated by their practical sense (Bourdieu, 1980) they have opposed a power that excludes or marginalises them from the social milieu to which they belong.

The Federation of Family Farm Workers (FETRAF) in Southern Brazil: a new unionism under construction

The third and last study discussed here focused on unionisation among family farmers in southern Brazil, as expressed in the 2001 creation of the FETRAF – SUL (Federation of Family Farm Workers for the Southern Region), linked to the nationwide labour federation, the

CUT (Central Única de Trabalhadores do Brasil). The main research question addressed here is an analysis of *whether* and *how* this federation permitted the emergence of new organisational forms and labour participation and struggles in the rural milieu, including widening the scope of these struggles to include family farmers as participants in the process of building a society based on solidarity and sustainability.

Research began with the analysis of the discourse and practices of the new federation, as elaborated individually and collectively by the social actors participating in the field of unionism[10] (union leaders, family farmers) and with the examination of the viewpoints of other actors participating in the wider organisational field of family farming – regional, micro-regional and local leadership.

The FETRAF came into being as a sort of inheritor of the movement of rural workers and small family farmers who opposed the type of unionism that proliferated after the 1964 military coup and underwent further expansion in the 1980s. The founders of this new federation were trained in the heat of the struggle against "state assistance" unionism organised through centralised federative structures that were weakly committed to the idea of the participatory representation of farmers and rural workers.

Thus, the federation was set up with the aim of widening the spaces of participation for farmers and integrating community and micro-regional instances into union organisation. With this founding principle as its thrust, the federation – in spite of its limitations and the still incipient character of its proposals – has fostered the emergence of innovative practices in the field of unionism.

Among these practices, we find their proposal for organisation at the work site (OLT). Although the national union federation CUT had already considered this type of proposal, it had not yet become widespread. In the case of the FETRAF, however, the practice has been implemented and it involves organisation at the level of rural communities.

Family farmers' participation at the various levels has permitted the constitution of more democratic spaces for discussion and debate. In this regard, the strategy implemented has been to work for the formation of community councils that are able to guarantee a more "capillary action" approach to union organisation. It is not enough to hear what farmers think or what they want for family farming; effective quality participation in the definition of the entity's struggles and strategies must be guaranteed. The grassroots farmer must feel capable of participating and be able to feel self-esteem as a farmer and citizen.

---

10 We are employing Bourdieu's notion of field (1983) here.

However, the community spaces that have been set up do not yet represent a consolidated and widespread phenomenon, and in particular unions and micro-regions show fragility in terms of mobilisation, struggle and effective participation on the part of leadership. Participation in executive management, for example, reveals the absence of leadership from particular micro-regions, specifically those that are more fragile in a political sense and thus remain on the periphery of union organisation.

The approach that has been used to reach out to rural communities has also been innovative, harking back to a traditional rural practice: the family farmers' *mutirão*.[11] Union members go into the diverse communities that the federation represents with the purpose of encouraging them to speak, discuss, propose and get involved in the organisation. This is not meant merely to identify spokespeople for the organisation, but to encourage genuine representation. The *mutirão* is a way of reinvoking the solidarity of neighbours for the fulfilment of the collective task of organisation, the building of a common project for which one needs to believe that the other can help, can contribute and must be valued. To value each member belonging to the field means recognising that each one has the right to "play the game", but as a full player and not as a mere witness or spectator.

Although still insufficiently implemented, the *mutirão* is meant to permit family farmers to *gain power*, building a process of *empowerment* based on the widening of participatory spaces that are no longer restricted to the "right" to delegate to others the powers of representation. Those who are represented gradually take on responsibility for being the spokespeople for their own desires, dreams and proposals, even when they go on to delegate their "voice" to someone else (union leadership). In the family farmers' own evaluation, the *mutirão* is recognised as valuable and worthy of further development and implementation.

There is another new factor contributed by FETRAF discourse and practices when it comes to representing family farmers: the perspective of not limiting action to a union-based struggle but also positioning themselves as protagonists in the construction of a social project for sustainable and solidarity-oriented development. In order to do so, they consider it necessary to act in consort with other social entities, particularly those that pertain to the organisation of family farming. In this dimension of their work, actions linked to production, seen as strategic

11 Translator's note: the term *mutirão* refers to a traditional popular practice in Brazil, a gathering of neighbours, relatives and friends to collectively and collaboratively realise a special task for which an exceptional deployment of labour is needed, such as putting up the walls of a house.

for the consolidation of farmers and federation, are given priority. For this purpose, partnerships have been built with other organisations and NGOs that act within the field of family agriculture, seeking to define practices that will be beneficial to farmers, with an emphasis on production, industrialisation and commercialisation, and privileging alternatives such as agro-ecology and economic solidarity networks. From this perspective, the federation also provides particular support for the organisation of farm women and youth, taking the view that any perspective on social change must include the recognition of the gender and generation-based cleavages in family farming.

The potential and the limitations for strengthening family farming through projects for sustainable and solidarity-oriented development has led the federation to consider partnerships within the field of family farming. This is not just in terms of an abstract principle of articulation but – above and beyond all else – in terms of real forms of solidarity, notwithstanding the conflicts and power disputes among the different organisations that make up this social field. The federation thus seeks to make viable the collective practices for formulating public policy demands and for farmers' collective consolidation as political actors in society.

In this regard, the FETRAF, after four years of existence, can be qualitatively distinguished from older forms of unionism, although it is still affected by the contradictions and historical limitations of union organisation. The old *habitus* (Bourdieu, 1983) of union leaders has been questioned and a new one has begun to be built, especially in the sense of opening up to more effective participation by the farmers themselves, which leads them to effective *power gains* within union structures. Lastly, the union's role in a larger project for sustainable and solidarity-oriented development is another FETRAF innovation intended to involve family farmers in a wider process of *empowerment*, one that is society-wide.

Conclusion

As we have already seen, the notion of *empowerment* is polysemic and serves diverse ends. The three experiences that have been studied here can be analysed under the perspective of community *empowerment*, but also the Pignarre and Stengers (2005) view that was presented in section 2. Thus, the notion of *empowerment* has been used as "technique", strategy and process for the constitution of sites where symmetric relations are built, to allow for the collective construction of alternatives to what is experienced as restrictive, excluding and unsatisfactory. Included in this sense is the change in subjectivities – a subject is built through the recognition of the other; furthermore, there is the notion that the process can place the group in a position to influence and participate effectively in the elaboration and application of public policies geared toward its needs and to become part of a wider process of social change geared toward the project/utopia of a sustainable society.

In all three cases, these are experiences in which the activity of local communities was the key force behind their constitution; in different ways, we see the creation or use of community spaces – neighbourhood groups, kinship groups – to favour discussion in symmetric conditions: the circle, and informality, prevail over the institutions and organisations involved in the origin of actions. Non-violence is the preferred course, but confrontation is not avoided; the struggle for social inclusion and for the smaller-scale dimensions of change are given precedence in a discourse that is constructed in opposition to hegemonic patterns of social relations: a discourse presenting a project for sustainable development in which remaining a family farmer is possible and valued (as exemplified by the Rede Ecovida and the FETRAF), a society in which there is room for small-scale agriculture, even where environmental legislation and its implicit idea of "inviolable nature" prevail (as with the APA in Guaraqueçaba).

Nor is it difficult to see what remains of hierarchical and asymmetrical internal relations in the groups studied here. Even if asymmetry is the rule and conflict is the basic dynamic of our society, experiences such as those studied here can still build alternative relational structures. To the extent that their practices are rooted in rural communities, in daily life, in conversation and discussion circles, in the exchanges and networks of farmers, there is a strong willingness to hear and to speak, to revalorise or build identities that favour *empowerment* in the creation of collective alternative strategies.

In these strategies of reaction and of valorisation of the identity of farmer and family farmer, and of care-takers of the environment – artifices of another approach for development – the process of revitalisation of the rural as a site for life and work is underway, all the while contributing to a broader movement to counter the dominant model of society.

References

Beaulieu M., *La sociologie de la santé et du vieillissement*, Palestra proferida na UFPR, Curitiba, SCS, 18 April 2005.

Becker D. *et al.* "Empowerment e avaliação participativa em um programa de desenvolvimento local e promoção da saúde", *Ciência e saúde coletiva*, 9, 3, Rio de Janeiro, July 2004, pp. 655-667

Bourdieu P., *Le sens pratique*, Paris, Minuit, 1980.

Bourdieu P., *Questões de sociologia*, Rio de Janeiro, Marco Zero, 1983.

Bourdieu P., *Coisas ditas*, São Paulo, Brasiliense, 1987.

Bourdieu P., *Razões práticas: sobre a teoria da ação*, Campinas, Papirus, 1997.

Bourdieu P., *O poder Simbólico*, Rio de Janeiro, Bertrand Brasil, 1998.

Bourdieu P., *O Campo Econômico: a dimensão simbólica da dominação*, Campinas, Papirus, 1999.

Bourdieu P., *La distinción: critérios y bases sociales del gusto*, 2nd. ed., Madrid, Taurus, 2000.

Bourdieu P., *Meditações pascalianas*, Rio de Janeiro, Bertrand Brasil, 2001.

Bourdieu P., *Les structures sociales de l'économie*, Paris, Seuil, 2002.

Brandemburg A., *Agricultura familiar, ONGs e desenvolvimento sustentável*, Curitiba, Editora da UFPR, 1999.

Brandemburg A., *Sociologia do ambiente rural: principais temas e perspectivas*, Indaiatuba/SP, ANPPAS, 2003.

Breton M., "Relating Competence: Promotion and Empowerment", *Journal of Progressive Human Services*, 5, 1, 1984, pp. 27-45.

Carvalho S.R., "Os múltiplos sentidos da categoria *empowerment* no projeto de Promoção à Saúde", *Cadernos de Saúde Pública*, 20, 4. Rio de Janeiro, FIOCRUZ, July/August 2004, pp. 1088-1095.

Ferreira Duarte Damasceno A., *O pensamento social sobre o rural e o urbano*, Curitiba, UFPR, 2000, CD-Rom.

Ferreira Duarte Damasceno A., "Vers la reconstruction du rural au Brésil", in Zanoni M., Lamarche H. (eds.), *Agriculture et ruralité au Brésil*, Paris, Karthala, 2001.

Ferreira A., "Processos e sentidos sociais do rural na contemporaneidade: indagações sobre algumas especificidades brasileiras", *Revista Estudos Sociedade e Agricultura*, 18, Rio de Janeiro, CPDA/UFRRJ, April 2002, pp. 28-46.

Ferreira A. *et al.* "Resistência e Empoderamento no meio rural", *Revista Estudos Sociedade e Agricultura*, 28, Rio de Janeiro, CPDA/UFRRJ, December 2007, pp. 123-159.

Ferreira A. *et al.*, "Les stratégies des acteurs locaux vers la construction d'un développement territorial durable et solidaire dans le contexte de la mondialisation et des changements globaux", *Communication présentée au Colloque de l'ASRDLF*, Quebec, UQUAR, août 2008.

FETRAF-SUL/CUT, *Resoluções do 1º Congresso Sindical da Agricultura Familiar*, Chapecó, FETRAF-SUL, março 2001.

FETRAF-SUL/CUT, *Agricultura familiar, desenvolvimento e o novo sindicalismo: da vida que vem da terra, a semente de um novo Brasil... semente do novo sindicalismo*, Chapecó, FETRAF-SUL, March 2002.

FETRAF-SUL/CUT, *Mutirão da agricultura familiar*, Chapecó, FETRAF-SUL, December 2003.

Fisher, R., Kling J, "Community organization and new social movement theory", *Journal of Progressive Human Services*, 5, 2, 1994, pp. 5-23.

Floriani D., *Conhecimento, meio ambiente e globalização*, Curitiba, January 2004.

Gorz A., *Metamorfoses do trabalho: crítica da razão econômica*, São Paulo, Anna Blume, 2003.

Hamilton D., "O revivescimento da aprendizagem?", *Educação e Sociedade*, 23, 78, 2002, pp. 187-198.

Herrick J., "Empowerment Practice and Social Change: the Place for New Social Movement Theory", *Proceeding "The New Social Movement and Community Organizing Conference"*, Seattle, WA, University of Washington, 1995.

Jollivet M. (ed.), *Vers un rural postindustriel; rural et environnement dans huit pays européens*, Paris, L'Harmattan, 1997.

Jollivet M., Pavé A., "O meio ambiente: questões e perspectivas para a pesquisa", in Vierira P.F., Weber J. (eds.), *Questão de recursos naturais renováveis e desenvolvimento: novos desafios para a pesquisa ambiental*, São Paulo, CORTEZ, 1997.

Jurema Bentes S., "Ações e estratégias do CNDM para o 'empoderamento' das mulheres", *Rev. Estud. Fem.*, 9, 1, 2001, pp. 207-212.

Lamarche H. (ed.), *A agricultura familiar : comparação internacional, volume I : uma realidade multiforme*, 2nd. ed., Campinas, Editora da UNICAMP, 1997.

Lamarche H. (ed.), *A agricultura familiar : comparação internacional. volume II : do mito a realidade*, Campinas, Editora da UNICAMP, 1998.

Lamarche H., *Rural et agriculture ou l'histoire d'un impossible divorce*, Working Paper, Paris, Ladyss-CNRS, 2000.

Luginbuhl Y., "Le rural pour repenser la nature?" in Mathieu N., Jollivet M. (eds.), *Du rural à rural l'environnement : la question de la nature aujourd'hui*, Paris, L'Harmattan, 1989.

Maresca S., *Les dirigeants paysans*, Paris, Minuit, 1983.

Miguel L., Zanoni M., "Impactos socioeconômicos de uma política de proteção ambiental sobre os agricultores familiares e produtores rurais do litoral norte do Estado do Paraná", in Aguia D., Pinho J.B. (eds.), *Agronegócio Brasileiro: desafios e perspectivas, 36° Congresso Brasileiro de Economia e Sociologia Rural. Artigos e Resumos*, Brasília, SOBER, 1998.

Miguel L., Zanoni M., "Reflexões sobre a hostilidade das populações locais frente à implantação de políticas regionais de proteção ambiental no Brasil: o caso da A.P.A. de Guaraqueçaba – Paraná", *II Simpósio Internacional de Qualidade Ambiental. Anais...*, Porto Alegre, UFPR, 1998a.

Pignarre Ph., Stengers I., *La sorcellerie capitaliste*, Paris, La Découverte, 2005.

Pinheiro G. S. R., "Agricultor familiar e projeto agroecológico de vida", *Dissertação apresentada ao Mestrado em Sociologia*, Curitiba, UFPR, 2004.

Pinto C. "Empowerment: uma prática de serviço social", *Política Social*, Lisbon, ISCSP, 1998, pp. 247-264.

Raynaut C., "O desenvolvimento e as lógicas da mudança: a necessidade de uma abordagem holística", *Os cadernos de Desenvolvimento e Meio Ambiente*, 1. Curitiba, Editora da UFPR, 1994.

Raynaut C., Zanoni M., "Meio ambiente e desenvolvimento: imperativos para a pesquisa e a formação. Reflexões em torno do doutorado da UFPR",

Cadernos de Desenvolvimento e Meio Ambiente: sociedade, desenvolvimento, meio ambiente, 1, Curitiba, UFPR, 1994.

Raynaut C. *et al.*, “Sustentabilidade e mudança : História das transformações sociais e ecológicas no litoral do Paraná (Brasil)”, *Anais do X World Congress of Rural Sociology/XXXVIII Congresso Brasileiro de Economia e Sociologia Rural*, Auburn/Al – Brasília/DF, IRSA/SOBER, 2000, pp. 1-27 (CD-Rom).

Riger S., “What’s Wrong with Empowerment”, *Community Psychol*, 21, 1993, pp. 279-92.

Rissel C., “Empowerment: the Holy Grail of Health Promotion?” *Health Promot Int*, 9, 1994, pp. 39-47.

Rodrigues A.S., “A federação dos trabalhadores na agricultura familiar: um novo sindicalismo em construção”, *Dissertação apresentada junto ao Mestrado em Sociologia*, Curitiba, UFPR, 2004.

Rodrigues C., Santos F., “Empowerment: ciclo de implementação, dimensões e tipologia”, *Gest. Prod.*, 8, 3, December 2001, pp. 237-249.

Santos E., “Terra de agricultura, terra de conservação: repensando a natureza, redefinindo o rural : o caso da comunidade rural de Batuva, município de Guaraqueçaba-Paraná”, *Dissertação apresentada junto ao Mestrado em Sociologia*, UFPR, 2004.

Touraine A., *O que é a democracia?*, Petrópolis, RJ, Vozes, 1996.

Vieira P., “Meio ambiente, desenvolvimento e planejamento”, *Meio ambiente, desenvolvimento e cidadania: desafios para as ciências sociais*, São Paulo, CORTEZ/UFSC, 1995.

Silva O., “Considerações sobre a monografía. Estratégias dos Agricultores familiares num APA”, in de Eduardo, B Santos (eds.), *Defendida como requisito para obtenção do grau de bacharel em Ciências Sociais*, Curitiba, UFPR, 2000.

Silva O., *Les réponses des organisations syndicales agricoles françaises à la question de l’agriculture durable*, working paper, Paris, Ladyss-CNRS, 2005.

Wallerstein N., “Powerlessness, Empowerment, and Health: Implications for Health Promotion Programs”, *Am J Health Promot*, 6, 1992, pp. 197-205.

Wanderley M., “Raízes históricas do campesinato brasileiro”, *XX encontro anual da ANPOCS. GT 17 Processos Sociais Agrários*, Caxambu/MG: ANPOCS, 1996.

Wanderley M., “Em busca da Modernidade Social: uma homenagem a Alexander V. Chayanov”, in Ferreira A.D.D. and Bradenburg A. (eds.), *Para pensar OUTRA AGRICULTURA*, Curitiba, UFPR, 1998.

Wanderley M., “Emergência de uma Nova Ruralidade nas Sociedades Modernas Avançadas: O ‘Rural’ como Espaço Singular e Ator Coletivo”, *Estudos Sociedade e Agricultura, outubro*, 15, 2000, pp. 87-146.

Wanderley M., “Territorialité et ruralité au Nord-est ; vers un pacte social pour le développement rural”, in Zanoni M., Lamarche H., *Agriculture et ruralité au Brésil*, Paris, Karthala, 2001.

Wanderley M., “Identidade social e espaço de vida”, *Texto apresentado para o VI congresso da Associação Latino Americana de Sociologia Rural (ALASRU)*, Porto Alegre, UFPR, 2002, CD-Rom.

Weissberg R., *The Politics of Empowerment*, Westport, Praeger Publishers; 1999.

Zanoni M. *et al.*, "Preservação da natureza e desenvolvimento rural: dilemas e estratégias dos agricultores em Áreas de Proteção Ambiental", *Desenvolvimento e Meio Ambiente : A reconstrução da ruralidade e a relação sociedade/natureza*, 3, Curitiba, UFPR, 2001.

Participatory Approaches in Policy-relevant Knowledge Production

Pieter LEROY

Professor, Radboud University Nijmegen, the Netherlands

In recent years "uncertainty" has become a key issue. The debate is taking place in sociology and political sciences, environmental studies, science and technology studies, as well as other disciplines and domains. Whatever its precise meaning, uncertainty refers to the fact that contemporary society faces increasingly complex problems that seem to be beyond the problem resolution and management capacities of modern institutions in the realms of politics, science and technology. While the focus of this contribution is on environmental issues, these problems and the uncertainties that result from them are not restricted to environmental issues. Environmental threats are only one part of the generalised social risks encompassing politics, economics and culture.

One of the indicators of uncertainty often referred to is the increasing lack of legitimacy of modern institutions in politics, science and technology. In accordance with Beck's "risk society" concept, Healy describes contemporary society as one in which "the condition of heightened urgency, risk and uncertainty as propelled by the uncontrollable technological innovation, has led to a loss of trust in both science and institutions as credible authorities to cope with stress" (Healy, 2001: 41). This (perceived) lack of steering capacity makes politics and science vulnerable, as they suffer from a loss of confidence. As Wynne (1996) pointed out, while this lack of legitimacy may often be latent and invisible, it is very active or even explosive at times. For example, when technological shortcomings become evident, as with nuclear waste; when science cannot solve or undo certain problems, as has been the case from BSE to GMOs; and when politicians appear unable to deploy effective and accepted strategies, for instance on climate change, on global inequality, and on the proliferation of nuclear arms. In these and similar cases, science, technology and politics seem to have lost credibility and authority, giving rise to a mix of distrust, protest and alternative problem-solving strategies.

In an attempt to regain ground, we have seen the emergence of participatory approaches everywhere. The almost simultaneous emergence of new forms of participatory governance and their spread in different policy domains and in politics in general, including technology assessment and knowledge production, cannot be mere coincidence. First, they seem a response to largely similar problems arising in these different spheres: a lack of steering capacity, performance and legitimacy, as evoked above. Second, these participatory approaches are advocated through largely similar arguments. Two arguments recur: participatory approaches, firstly, are said to contribute to better-informed decisions, and thereby to the quality of steering efforts, as they take on board stakeholders' perspectives. Participation, secondly, should contribute to more legitimate processes, as participation is presumed to foster acceptance. In brief, participatory approaches should enhance input-legitimacy as well as output-legitimacy. All these arguments are normative in nature, as they advocate a further step in the perpetual quest for (new forms of) democracy. In addition, these participatory approaches should help overcome the shortcomings of modern institutions, while canalising uncertainty and remedying lack of legitimacy.

In the meantime, the idea(l) of participatory governance has led to elaborate normative frameworks and to a range of innovative societal practices. This contribution is aimed at assessing what difference these participatory strategies and initiatives make. It does not attempt to give definitive answers as to how far participatory initiatives really lead to better-quality decision-making, more legitimacy and greater steering capacity. Rather, it positions these participatory initiatives within a more comprehensive view of recent shifts in (participatory) governance (section 2), and within different approaches to participatory knowledge production (section 3).

As mentioned, the present contribution focuses on environmental issues, more specifically on knowledge-extensive processes of environmental decision-making. In exploring theoretical and empirical research, it builds upon earlier writings on recent developments in environmental politics (Van Tatenhove, Arts and Leroy, 2000; Arts and Leroy, 2006), on the role of participation therein (Leroy, 2002; Van Tatenhove and Leroy, 2003), and on the production of policy-relevant environmental knowledge (Turnhout and Leroy, 2004). As such, this contribution, at the time of initial writing (2005), formed the basis for a project with the Netherlands Environmental Assessment Agency on the coproduction of knowledge in the context of environmental reporting and assessment, now completed (Hage and Leroy, 2007; Hage, Leroy and Petersen, 2010).

Governance and participatory approaches

Governance

Contemporary policy-making is increasingly referred to as "governance". There is no need to echo the different versions and connotations of this concept here (Rhodes, 1997; Van Kersbergen and Van Waarden, 2004; Kooiman, 2002), yet the flexible character of the concept is part of its success. To account for the widespread adoption of "governance" in general and "participatory governance" in particular, is to point to its multi-interpretability (Hajer, 1995).

These new modes of governance are said to challenge and alter traditional government strategies. First, governance is regarded as a way out of hierarchical intervention and the failures associated with top-down coordination; consequently, it represents more horizontal forms of steering: interaction and networking among parties, partnerships, self-governance and similar mechanisms. Second, governance implies a shift in the locus of politics: from constitutional politics to politics outside traditional polity frameworks and institutions, from national to both sub-national and supranational levels (Van Tatenhove and Leroy, 2003). Yet these developments do not imply the abolishment of "government" (Pierre, 2000); rather, they represent the increasing juxtaposition of government and governance practices. Governance thus refers to a gradual transformation of the political domain where the formulation of goals, the choice of instruments and the implementation of solutions becomes the combined task of political actors, corporate interests, civil society and transnational organisations. Some scholars suggest that the state has become an adaptive entrepreneur that performs several roles: it commands, controls, regulates and executes, while it increasingly has to deal with the complexity and dynamics of modern societies, and therefore increasingly interacts, cooperates, coordinates and facilitates. It demands "governmentality", a political style that can best be described by the principles of "reflexive rationality" (Theys, 2002; Jessop, 1998). This includes the ability to respond in a creative and flexible manner to problems of governance, a willingness to learn on practical and institutional levels, and the stimulation of normative debates on the principles underlying all governance activities (Kooiman, 2000 and 2002).

Participatory governance on environmental issues

The quest for this reflexive governmentality has resulted in a variety of participatory strategies and projects, among others in the environmental field. Despite their differences in scale, issue, design, stakeholders involved, outcome and level of success, we label them all "participatory governance approaches" here, while restricting our focus to environmental issues. This section develops a typology of analyses as these emerge from the literature. Section 2.3 will survey the ways these participatory governance approaches have been assessed with regard to their added value on capacity and legitimacy issues.

While the environmental policy domain was initially characterised by hierarchical strategies of intervention and regulation, it has recently played an *avant-garde* role in the innovation of forms of governance (Theys, 2002). The multitude of innovative forms of environmental governance is often captured by terms such as "deliberative", "discursive" or "participatory". Many of these innovative projects have been investigated in a series of case studies and by a series of scholars, resulting in a vast number of publications (see among many others, Healey, 1997 and 1998; Pløger, 2001; Keller and Poferl, 2000; Sharp and Richardson, 2001; Yearley *et al.*, 2003; Flyvbjerg, 1998; Smith and Wales, 2000; Irvin and Stansbury, 2004; Spash, 2001). With regard to France, we refer, on a theoretical level, to Callon *et al.* (2001) and, for a series of empirical studies, to the publications from the research programme "Concertation, décision et environnement" (see http://concertation-environnement.fr), covering different subfields of the environmental domain.

Table 1 summarises this literature by classifying the sorts of analysis on participatory environmental policy-making – in modern industrialised countries – while distinguishing these around two axes:
– the level of analysis, ranging from a mainly conceptual towards a more empirical approach, and
– the mode of analysis, ranging from a descriptive to a prescriptive or normative approach.

By crossing these axes, we get four boxes that indicate a path through the existing literature, even though many authors and contributions cannot be captured by reference to a single box. In addition, the axes refer to an analytical continuum, rather than to a material classification.

The upper boxes comprise empirical approaches, *i.e.* analyses that focus on specific participatory practices, either presenting them as such (how it *is* done), or commenting on them in a normative way (how it *should be* done). The latter approach is usually aimed at prescribing what instruments could or should be used and what criteria should be fulfilled to enhance the quality and legitimacy of decision-making (Joss and Bellucci, 2002; Kasemir *et al.*, 2002). In more operational terms, it often leads to a number of recommendations or offers, a toolbox for more legitimate and more effective processes of participation.

Mode of analysis / Level of analysis	***Descriptive***	***Prescriptive (normative)***
Empirical	Report on experiences with participatory practices and procedures in environmental governance, on local, national or supranational level	Formulate suggestions, recommendations for practices. Provide a toolbox with "how to" techniques for participation
Conceptual	Characterise and explain trends in governance. Consider participatory governance within broader processes of socio-political change	Provide normative (meta) principles for participatory governance. Generally according to the Habermasian ideal of "communicative rationality"

Table 1 – Four types of analysis of participatory environmental governance

The lower boxes refer to more conceptual approaches and usually describe participatory governance as a discursive concept. These analyses go beyond specific practices and try to identify the characteristics of governance responses to complex problem-solving in modern societies. Kooiman (2002) provides one of the most elaborated accounts on governance in general. He distinguishes between first-order, second-order and meta-governance. First-order governance comprises the "day-to-day activity of public and private actors in concrete governing situations" (*ibidem*: 86). Second-order governance focuses on the design of adequate institutional structures, which should precede and enable participatory problem-solving by establishing a set of rules, instruments and resources. Third-order or meta-governance concerns "the governing activities aimed at the broad principles that concern the way governance itself, either first or second-order, takes place" (*ibidem).*

Most empirical studies (upper boxes) deal with first and second-order governance, the former referring to the projects, the latter to their institutional context. When authors explicitly engage with third-order or meta-governance, they may formulate the very principles of participatory governance, in either an analytical or a prescriptive way (lower boxes). For instance, Keohane is preoccupied with the building of appropriate institutions for global governance that are both legitimate and effective fora for the negotiation of policies. He thereby engages with a normative analysis of global institutions based on a Habermasian line of thought that outlines how "voluntary cooperation based on honest communication and rational persuasion provides the strongest guarantee of a legitimate process" (Keohane, 2002: 263).

In summary, two core characteristics emerge. First, the literature on participatory approaches commonly refers to a democratic or legitimacy deficit, as forms of classical, representative accountability no longer suffice in the age of complexity (Kooiman, 2000). For many authors, participation has become an imperative in a context of "deep uncertainty" (Pellizoni, 2001, 2003a and 2003b), "unstructured environmental policy problems" (Hisschemöller and Hoppe, 1996; Bulkeley and Mol, 2003) and a general "decline of governability" (Kooiman, 2002, Theys, 2002). Environmental issues in particular are challenging, as these issues are less and less structured, and more and more complex in nature (Blowers and Leroy, 1996; Held *et al.*, 1999). In other words, these scholars refer to the (lacking) capacity and legitimacy, referred to in the introduction. Participatory governance is assumed to be the remedying strategy to overcome these problems. Therefore, participatory governance or "deliberative democracy" (Hajer and Wagenaar, 2003; Pellizoni, 2003a and 2003b) is the most desirable way to tackle the deficits of traditional polity and policy.

Second, the literature on participatory approaches is primarily concerned with questions of legitimacy, and with an improved level of effectiveness through greater legitimacy. As such, its strategy clearly differs from problem-solving strategies based on neo-liberalism and market logic that transfer the criteria of efficiency pertaining to economics onto politics and policy domains. Having said that, the literature on participatory approaches is ambivalent with regard to the emergence of public-private and private-private partnerships as a means of problem-solving. The actual contribution of partnerships and the circumstances under which they can fulfil the ambitions they have been ascribed to, have only recently been investigated empirically (Glasbergen *et al.*, 2007; Visseren-Hamakers, 2009; Van Huijstee, 2010).

Participatory environmental governance: from reformist suggestions to radical pessimism

Our classification of the literature reveals that most authors have a normative commitment to democratising environmental governance, as they signal a need to move away from hierarchical or market-led governance, and sketch the outlines for new deliberative modes of governance. Recently, though, we have seen a growing number of both empirically and theoretically-minded scholars critically assessing these new forms of governance in relation to citizenship, to legitimacy and to the wider socio-political context, paying attention to the possible failures of participatory governance, either in general (Jessop, 1998 and 2000) or with regard to environmental issues.

These critical examinations, however, arrive at different conclusions, depending on the empirical examples and the area of study and, of course, the criteria used. Consequently, there is no consensus on the crucial question of whether the experiments with participatory governance organised hitherto should be assessed as successful and effective, or as ineffective and wasteful (Irvin and Stansbury, 2004). Given the variety in both empirical cases and criteria used, we restrict here to a typology of the criticisms raised.

In building such a typology, it is easy to distinguish between approaches that emphasise operational, organisational and procedural aspects of participatory initiatives, and those that consider the institutional and political context to be crucial for effective participation. Thus, assessments of participatory governance arrangements depend upon two distinguishable evaluative considerations:

– a (positive or negative) assessment of the quality of the *participatory design*, including questions such as: to what extent are the participatory mechanisms able to engender inclusion, mutual understanding and cooperation? To what extent do they connect to the problem at hand? To what extent is the participatory design (consensus meeting, referendum, public debate, citizen juries and others) suited to the issue at stake and under the given circumstances?

– a (positive or negative) assessment of the *contextual conditions*, including questions such as to what extent the pre-existing contextual factors (polity institutions, rules of the game, power interrelations, etc.) constrain effective implementation or generally complicate the application of innovative participatory approaches.

The combined assessment of participatory design and conditions of implementation yields Table 2. Again, the four boxes do not represent absolute stances. In fact, scholars take highly nuanced positions in the debate and tend to consider both weaknesses and strengths of both design and context. Nevertheless, this typology, for the sake of clarity, groups together authors and assessments, with either an optimistic view (upper left box) or a radically pessimistic one (bottom right box).

Quality of participatory design / Conditions of implementation:	+	-
Moderate barriers to the implementation of participatory designs	I **"Right direction!"'** Participation processes function relatively well. Societal and political hurdles may have to be overcome, but relatively optimistic about possibility of implementing them	II **"Procedures"** Due attention to design of methodologies. Consequent 'right' application of discursive techniques. Assumes more or less favourable conditions for implementation
Substantial or structural (institutional, epistemological, power-related) barriers	III **"Politics"** Lack of political will to ratify deliberative approaches due to structural barriers. No clarity on status of deliberative processes in political decision-making.	IV **"Power"** Habermasian deliberative practices are naïve in a Foucauldian reality. Persistent power relations present structural barriers to a proper implementation and application of participatory designs.

Table 2 – Four critical approaches to participatory governance (adapted from Theys, 2002, pp. 234-236)

The "right" direction! (I)

Drawing upon the Habermasian ideal of "non-coercive communicative practices", Healy (1998) emerges as one of the advocates of "collaborative" or "deliberative" planning. Based on extensive study, she sees opportunities for commonly-motivated stakeholders to arrive at mutually-binding win-win situations through "inclusive interactive strategy building". Her optimistic view on collaborative planning is hardly tempered by institutional constraints, although "it is clear that the evolution of 'good practice' in collaborative planning is not just a matter of the capability and commitment of those involved in particular practices. Its possibility is encouraged or constrained by the institutional context" (Healy, 1998: 16), and "multi-stakeholder collaborative planning [...] is [...] severely constrained" (*ibidem*: 17). Nevertheless, she expects considerable transformations of the current system to take place.

Procedures (II)

A considerable amount of the literature focuses on the proper design of participatory processes and methodologies, with only moderate reference to contextual constraints. Convinced of the potential of these participatory procedures, these scholars aim to enhance the methodological quality of these participatory processes (Goodin and Niemeyer, 2003; Richards *et al.*, 2004). This involves the selection of participants, the solving of representational issues, the link between the deliberation process and constitutional decision-making, the timing of the discursive process, etc. – in short, effective process management (Holländer and Leroy, 2001). While questions of institutional context are raised (among others, Smith and Wales, 2000), this is not given a negative connotation.

Politics (III)

The positions taken in boxes III and IV differ fundamentally from those in the upper boxes in that they assume that structural barriers hinder real participation. These barriers may be institutional (e.g. constitutional procedures) or epistemological (expertocratic or technocratic forms of knowledge input); in brief: based upon traditional power-related politics. This results in a pessimistic view of the potential of participatory approaches to change pre-existing political interests.

Magnette (2003), for example, observed how EU-sponsored initiatives fail to encourage ordinary citizens to become more active. No groups other than those already well organised are likely to be able to take part in participatory governance. This is due to complex procedures of decision-making and institutional barriers, the tendency to dissolve political issues into technical ones, as well as the limited interpretation of participation as being some form of consultation.

Power (IV)

Whereas box III regards (the lack of) political will as the crucial hindering factor, Pløger (2001) and others go a step further, contrasting the Habermasian ideal with a Foucauldian "real life" view of power. Habermas assumes that the sincerity, comprehensibility, legitimacy and truthfulness of arguments are accepted as moral guidelines for "non-coerced reason", and therefore tends to leave out disagreement (very) and uneven power balances, etc. A Foucauldian analysis, however, emphasises institutional barriers, the power of planning rhetoric, political and economic interests that, in the best case, take advantage of participatory initiatives to disguise illegitimate power structures.

Participatory approaches to knowledge production

We now turn from participatory governance in general and related to environmental issues in particular towards the more specific issue of the participatory governance of policy-relevant knowledge. Not only have traditional practices of policy-making and government been discredited by a lack of capacity and legitimacy. Scholars argue that uncertainty itself represents a fundamental epistemological criticism (Irwin, 1995; Pearson, 1998). One of the tacit assumptions of policy-making in modern societies has been the idea that uncontroversial science and expert knowledge underpin policy-making processes. Yet with the emergence of uncertainty – represented by crises over BSE, GMOs, bird flu, UMTS and others – science itself suffers from a lack of legitimacy and can no longer be regarded as the sole source of legitimate knowledge (Torgerson, 2003; Taylor, 2001). With complexity and uncertainty, we witness the limits of traditional ways of knowledge production and of science's intimate and uncritical relationship with decision-making (Gibbons *et al.*, 1994; Nowotny *et al.*, 2001). Therefore, in a debate that has largely paralleled the discussion on governance, scholars have reformulated the principles and practices of scientific knowledge production and its interrelation with policy-making.

The science-policy interface

Knowledge is a crucial resource for policy-making. As a consequence, the production and utilisation of usable knowledge is a prime concern, in particular in modern, expert-driven western societies. The classical, so-called "two communities" view of the relationship between science and politics suggested a clear demarcation of tasks: policy-makers were supposed to ask experts for useful information, to which experts would respond with valid, reliable and usable knowledge that policy-makers in turn could build upon. The adage "speaking truth to power (Wildavsky, 1979) to express the role of scientists reflects the different communities scientists and politicians were assumed to work in, with quite different ambitions and goals, different driving forces and rationales, different responsibilities and different systems of quality control.

From the early 1980s Gieryn elaborated the concept of "boundary work" (Gieryn, 1983). Instead of maintaining a clear demarcation between them, the boundaries between science and knowledge are permanently established and blurred, named and redefined, claimed and ignored. While it is true that science and politics are different worlds, with different values and standards, their boundaries are neither principal nor given, but socially constructed and contingent. Boundary objects (concepts, problem definitions, models, standards, etc.), boundary workers (experts, scientific advisers) and boundary institutions (advisory boards, scientific committees) play a pivotal role in an ongoing process of construction, deconstruction and reconstruction of the boundaries between science and policy. The environmental domain is a clear example. Right from its very emergence, scientists and experts predominantly influenced environmental policy, providing the analyses, the models, the standards, the goals and the strategies. Scientists thus clearly crossed the assumed boundary with politics and affected the latter – in many cases without being held responsible. Environmental policies therefore clearly reveal the actual intertwining of science and policy-making. Gieryn's amendments to the two communities' metaphor provoked a series of empirical questions on how boundary work actually works, how it combines the (often contradictory) claims and demands from either side, and how boundary workers or science-policy entrepreneurs do their job.

Whether it was inspired by Gieryn's concept or not, the environmental policy domain gradually became the example *par excellence* of questions on the science-policy interface. Scholars from Science and Technology Studies, from Policy Sciences and from Environmental Studies increasingly asked two strongly interrelated main questions. First, as the actual impact of scientists and experts grew, questions emerged as to "who is actually speaking to whom?". In *The Fifth Branch*, Jasanoff (1990) pointed out the important but largely invisible and uncontrollable role of experts and advisers – as boundary workers. The functioning of expertise was increasingly questioned (Irwin, 1995; Wynne, 1996; Roqueplo, 1997) and other questions were gradually brought in: on the claims of science itself and its monopoly as the exclusive provider of "truth". Political questions on power and influence thus metamorphosed into epistemological questions. Second, as environmental scientists increasingly faced complex issues and uncertainties, questions arose as to "what is the quality of the scientific knowledge that our policies are based upon?". Issues such as nuclear energy (in the 1970s), acid rain (in the 1980s) and climate change (from the 1990s onwards) were emblematic for this critical questioning, through which epistemological and methodological questions gradually metamorphosed into political ones.

In other words, traditional forms of knowledge production and utilisation in policy-making were increasingly questioned in terms of quality, capacity and legitimacy. The next section sketches some main streams of thought and their suggestions as to how to overcome these deficits with, once again, more participatory approaches, in this case related to modes of knowledge production.

Knowledge for policy: a quest for participatory production practices

The characteristics of many environmental problems challenge classical scientific knowledge – and its relation to politics. Nuclear energy, climate change and biodiversity are: (1) highly complex issues, demanding unusual multidisciplinary cooperation; (2) global issues, needing to be analysed in their (unequal) consequences all over the globe; (3) long-term issues, presuming encompassing and lasting systems of monitoring; (4) issues that include irreducible uncertainties, requiring application of the precautionary principle; (5) issues that carry high social, economic and political stakes; and (6) issues that cut across traditional distinctions, thus calling for a systematic dialogue between science, society and politics (Blowers and Leroy, 1996).

There is an overwhelming amount of literature on questions such as how to deal with these characteristics, both in terms of knowledge production and in political terms. This section restricts itself to three main approaches, starting from an epistemological, an organisational and a normative point of view respectively. Yet this restricted sample is largely representative of the debate. The main issues at stake are: the production of knowledge that acknowledges its own limits and uncertainties; the organisation of more applicable knowledge; the inclusion of non-scientist expertise; the quest for legitimate knowledge; and the quality control of co-produced knowledge.

Post-normal science

Funtowicz and Ravetz (1992 and 1993) characterise environmental issues as problems in which "facts are uncertain, values are in dispute, stakes are high and decisions may be urgent". Facing such issues, policy-makers urgently request valid and reliable knowledge – "solving the scientific puzzle" – that science is not able to produce, being in a situation of "soft facts and hard values". In such circumstances of intrinsic scientific uncertainties and high political stakes, the demand goes beyond the capacities of "normal science" and of "applied science". Therefore, classical science needs to be complemented by other ways of understanding, by other forms of knowledge.

Funtowicz and Ravetz developed the concept of "post-normal science", which can be characterised as science where the Cartesian distinction between facts and values cannot be maintained. Under conditions of soft facts, hard value-related decisions must be made. This requires non-scientific, more precise, beyond-scientific or post-normal knowledge and methods. As traditional methods of scientific quality assurance do not work under these circumstances, quality assurance is one of the major challenges of post-normal science. Funtowicz and Ravetz suggest "extended peer communities", that "deploy 'extended facts' and take an active part in the solution of their problems" (Ravetz, 1999: 647). All the different stakeholders in the policy process can take part in these extended peer reviews, contributing their so-called non-expert knowledge.

Post-normal science thus clearly includes the idea of involvement of non-scientists and calls for a participatory approach to enhance the quality, relevance and legitimacy of the knowledge produced. At the same time, Funtowic and Ravetz do not offer operational indications as to how to organise post-normal science, how to assess and ensure its quality, etc. (Van de Kerkhof and Leroy, 2000). Their main concern is with qualifying an epistemological issue, rather than with providing an organisational or practical toolbox to respond to it. In terms of Table 1 (see above), they provide a primarily prescriptive framework of analysis and design, without formulating clear-cut suggestions about how to process and to organise.

Mode II knowledge production

Gibbons *et al.* (1994) launched the idea of a "new mode of knowledge production", namely "Mode II knowledge production", looking at the science-policy interface from a more organisational perspective. Biochemistry, computing science, life sciences and other fields they cited as examples of Mode II seemed to have more or less similar modes of knowledge production, which clearly contrasted with the previous ones (Mode I). The earlier mode of knowledge production was – and still is – mainly monodisciplinary, institutionalised in universities or research institutes, steered by rather inflexible long-term programmes, controlled in a hierarchical way, and largely academic in substance and output. Mode II knowledge production, in contrast, is defined as multi or even transdisciplinary, referring to the involvement of non-scientist actors, generated in a context of application, produced in a diversity of sites, in ephemeral or even virtual networks, highly flexible and reflexive, and steered by novel forms of quality control (Nowotny *et al.*, 2001).

The latter point recalls the extended peer review advocated by Funtowicz and Ravetz. The argumentation also runs parallel, as Gibbons and colleagues claim that the real quality test for scientific knowledge lies outside the laboratory, and includes the societal and political approval of findings, results and their implementation. In addition to testing its validity and reliability, knowledge should be tested on its "social robustness". This concept, once again, implies a procedural turn in scientific quality assessment, as it includes the involvement of non-scientists, representing civil society.

Though Gibbons and colleagues do not provide operational organisational proposals either, it is clear that the idea of Mode II inspired the organisation of a series of recent research programmes, particularly in the environmental domain. National Science Foundations increasingly invite societal partners to take part in design, assessment and societal validation of such programmes. Certain German-speaking European countries have initiated research programmes along the ideas of Mode II, albeit often labelled "transdisciplinary" (Thompson Klein *et al.*, 2001).

Social learning for sustainability

Whereas both the post-normal and the Mode II approaches deliberately include values, the "social learning" approach to knowledge production is even more explicitly normative. This approach is related to the emerging field of "sustainability science" (Kates *et al.*, 2001; Kasemir *et al.*, 2003). Its key claim is that joint knowledge production and mutual learning between science and society is necessary to foster the transition to a sustainable society, as it is assumed that the combination of different ways of knowing and learning will enable social actors to work in concert, even under conditions of uncertainty.

The chief motivation to include these non-scientific forms of knowledge is threefold. The first is linked to the multi-scale structure of global social-environmental systems, and the second to the multiple, interactive and cumulative character of environmental stresses. These characteristics, labelled "complexity" and/or "multidisciplinarity" by others, require a broad knowledge base. The third argument is linked to the governance approaches discussed earlier on: stakeholder involvement is required to foster commitment to the process of sustainable development (Siebenhüner, 2003). More than the others, the social learning approach relates to a political process, to raising environmental awareness, empowerment and societal change.

Participatory environmental knowledge production: some cases

Despite the variety of perspectives on new modes of knowledge production, the approaches discussed above do reflect great convergence. First, with respect to their analysis, they all question the exclusive claim by classical sciences of providing "true" knowledge, and they all refer to epistemological problems related to the complexity of contemporary societal issues. Second, with respect to their strategies, they all call for new forms of knowledge production, including the involvement of non-scientists in its quality assessment. In terms of Table 1 above, however, these approaches mainly address conceptual issues, failing to provide many operational suggestions. In terms of Table 2, these approaches reflect quite a positive attitude, assuming more or less automatically that participatory approaches will indeed further the quality and the impact of the knowledge produced.

Both characteristics require these rather conceptual approaches to be complemented by empirical findings. Fortunately, there is a growing literature on a range of experiments with participatory knowledge production. First, this literature reports on a huge variety of methods and processes: consensus conferences, citizen juries, future search conferences, scenario-development, back-casting and foresight procedures, participatory (local) planning methods, participatory product development, participatory modelling, etc. Second, this great variety of methods and processes has been applied on different scales (from local to national), differing in scopes, on varying issues, with different aims and within varying policy contexts. Third, even though it addresses rather similar ambitions and processes, the literature differs in the concepts and the criteria for assessment used, depending partly on theoretical stances and partly on the area of application.

The small sample of projects that we, very briefly, report on below is, by no means, representative of the multitude of cases described. Yet we hope to draw some general conclusions from them, in an attempt to build a framework for interpretation and assessment.

ULYSSES

One of the first citizen participation projects in Integrated Assessment was the ULYSSES project (Urban Lifestyles, Sustainability and Integrated Environmental Assessment), (Kasemir, Jaeger *et al.*, 2003). Its main ambition was to develop a new methodology of communication between practitioners, on the one hand, and users of environmental science, particularly citizens, on the other – based on the assumption that the public and policy makers can provide information that is useful for environmental modelling as a method to support decision-making. The project focused on different ways of framing climate change and, as a consequence, modelling it. Specially designed focus groups in seven European metropolitan areas were to ascertain citizens' views on climate change to interact with IA computer models. Other sessions involved regional decision-makers and representatives of the financial world and the media.

The project resulted in a detailed inventory of the way citizens perceive and frame climate change. The project has been a milestone in the development of a methodology for citizens' participation in Integrated Assessment. Moderators of the focus groups avoided using (implicitly framing) phrases such as "global warming". The moderators were instructed not to limit the discussion to the rationale of the models used and to avoid an expert role. One conclusion was that citizens tend to frame climate change in ethical terms, despite the scientific uncertainties. The scientists, in turn, learned about the usefulness of their models in this context. While Ravetz was very positive, claiming this was essentially post-normal, others are more critical, as the original idea of creating systematic feedback into Integrated Assessment modelling has proved difficult (Siebenhüner, 2004).

COOL

Another extensively described project on participatory Integrated Assessment was the Dutch COOL-project (Climate Options On the Long term), (Van de Kerkhof, 2004). In contrast to ULYSSES, this project did not involve ordinary citizens, but professionals representing a range of stakeholders. The overall aim of the COOL-project was to develop possible scenarios for long-term climate policies at national, European and global levels. The national part of the project was aimed at developing somewhat more concrete CO_2-reduction strategies within certain economic sectors, aiming at a 80% reduction of CO_2-emissions by 2050. Different sector groups used back-casting methodologies to identify pathways and options.

The COOL project clearly reflects a problem of many participatory projects, not only in knowledge production: the difficult trade-off between democratic ambitions on the one hand, and ambitions in terms of the quality of the knowledge acquired on the other. Previously setting the (hypothetical) goal of 80% emission reduction fostered a creative, efficient and well-focused process among the stakeholders involved, but prevented them from framing the problem and the goal-setting differently – as they were invited to do in ULYSSES.

In addition, both ULYSSES and COOL are weakly linked to political decision-making, as both represent largely noncommittal processes that did not affect actual policy-making. Although both ULYSSES and COOL were aimed at formulating political recommendations, decision-makers did not take part in either process, which may have increased the noncommittal character; yet both had a clear methodological emphasis.

Local environmental monitoring

While most projects on participatory environmental knowledge production are directed towards the stages of problem definition, framing, scenario-building and policy design, only a few relate to other stages. Yearley *et al.* (2003) report on a participatory modelling exercise that relates to monitoring at city level. The aim was to produce spatial representations of local perception and knowledge on air pollution. The project, carried out in three English cities, made use of a "community planning" method that works on group discussions by mapping.

The results of this three-city case study were encouraging on the usability of this kind of participatory knowledge production. First, the high degree of overlap of citizens' local air quality perceptions with the scientific models confirms the citizen maps as an accurate representation of air quality. As a consequence, the processes resulted in mutual trust of the (scientific and lay knowledge) information provided. Second, the citizens' maps were useful to locate sites for additional monitoring, precisely in those areas where the citizen expertise diverged from the scientifically modelled maps. Therefore, this seems to be one of the rare success stories of participatory knowledge production: the participatory modelling exercise produced new knowledge that appeared not only to be relevant to the policy-making process, but that actually was used therein.

National environmental reporting in Flanders

For some time now, OECD and EU-member states have been required to make a "state of the environment report" (SoER), usually on an annual or biennial basis. Over the years the format of these environmental reports has been largely harmonised. In most states, the SoER is mainly, if not solely, produced by public authorities, public research institutes and other data providers and experts.

The processing of the Flemish SoER, however, can be labelled as a new form of knowledge production. First, while making the Flemish SoER is the responsibility of the Flemish Environmental Agency (FEA), the process is guided by a steering committee that, apart from representatives of governmental bodies, also comprises representatives of employers' and employees' organisations, environmental action groups and some "independent" experts. Within the framework and format that is agreed upon internationally, the steering committee can emphasise certain environmental issues, giving less priority to others, introduce new issues, call upon contradicting expertise, etc. As a result, the report not only reflects scientific priorities, but also the agenda of the societal groups represented. Second, although a task force within the FEA has the final editing responsibility for the report, various authors are invited to contribute to the report's different chapters and to comment on draft versions. These authors and reviewers come from a variety of backgrounds: academia, interest groups, research institutes, private consultancy, industries, environmental organisations, etc. Over the years, this procedure has resulted in a real mobilisation of expertise. Moreover, by its very mobilisation and its system of "extended review", this process contributes to the increasing quality and legitimacy of the SoER, the product itself. In other words, the joint production of environmental knowledge has led to a report of high quality that is widely accepted and socially supported.

Conclusion

This contribution has reported on the literature on participatory governance in general (section 2) and on participatory knowledge production in particular (section 3). As stated, its aim was to endorse a project on participatory environmental reporting and assessment within the Netherlands Environmental Assessment Agency, a project that has now been completed. A review of this literature points to four groups of conclusions.

Both the debates on participatory governance and on participatory knowledge production reflect enthusiasm, based on largely similar arguments: participatory approaches compensate for the decreasing legitimacy both of politics and of scientific knowledge. A large part of this enthusiasm, however, builds on prescriptive perspectives, rather than on thorough empirical investigation. In terms of Table 1: the right hand boxes are relatively overpopulated.

While there is an overwhelming amount of empirical research, this research clearly emphasises issues of design, methods and management. This has led to a series of design recommendations addressing issues such as stakeholder selection, timing, methods, etc. In terms of Table 2: the upper boxes are relatively overpopulated.

Gradually though, there has been an increasingly critical assessment of participatory approaches, addressing mainly the actual quality and the impact of these participatory processes within a political and institutional context.

Regarding the former, critics question the value added by participatory approaches to the quality of knowledge in decision-making (Irvin and Stansbury, 2004; Rayner, 2003). These scholars question the often contradictory goals of these participatory approaches: open access for all manner of participants on the one hand, and ensuring higher quality knowledge production on the other. This leads Collins and Evans (2002) to the argument that the problem of legitimacy has been replaced by the problem of extension, as if participation – primarily meant to increase legitimacy – in itself could guarantee better quality automatically. Collins and Evans therefore suggest doing away with the absolute openness of participatory processes, replacing it by a selective, still expert-driven accessibility.

With regard to the latter, there is increasing questioning of the actual added value of these processes and of the cost-effectiveness of the commitment of resources often involved. According to the lower boxes of Table 2, doubts are arising as to the relevance of particular institutional and political aspects for processes both of knowledge production and decision-making. The feeling is that the established scientific-institutional context may hinder the formulation and implementation of new participatory approaches in knowledge production. Hitherto, the literature on participatory governance has tended to neglect the fundamental underlying power issues.

References

Arts B., Leroy P. (eds.), *Institutional Dynamics in Environmental Governance*, The Netherlands, Springer, 2006.

Beck U., *Risikogesellschaft – Auf den Weg in eine andere Moderne*, Frankfurt am Main, Suhrkamp, 1986.

Blowers A., Leroy P., "Environment and Society – Shaping the Future", Blowers A., Glasbergen P. (eds.), *Environmental Policy in an International Context, Volume 3: Prospects for Environmental Change*, London, Arnold, 1996, pp. 255-283.

Bulkeley H., Mol A.P.J, "Participation and Environmental Governance", *Environmental Values*, special issue, 12, 2003, pp. 143-154.

Callon M., Lascoumes P., Barthe Y., *Agir dans un monde incertain*, Paris, Seuil, 2001.

Collins H. M., Evans R., "The Third Wave of Science Studies: Studies of Expertise and Experience", *Social Studies of Science*, 32, 2002, pp. 235-296.

Flyvbjerg, B., *Rationality and Power – Democracy in Practice*, University of Chicago Press, 1998.

Funtowicz, S., Ravetz J., "Science for the Post-normal Age", *Futures*, 25, 1993, pp. 739-755.

Gibbons, M., H. Nowotny H., Limoges C., Schwartzman S., Scott P., Trow M., *The New Production of Knowledge*, London, Sage, 1994.

Gieryn T., "Boundary Work and the Demarcation of Science from Non-science", *American Sociological Review*, 48, 1983, pp. 781-795.

Glasbergen P., Biermann F., Mol A.P.J. (eds.), *Partnerships, Governance and Sustainable Development – Reflections on Theory and Practice*, Cheltenham, Edward Elgar, 2007.

Goodin R.E., Niemeyer S.J., "When Does Deliberation Begin? Internal Reflection versus Public Discussion in Deliberative Democracy", *Political Studies*, 51, 2003, pp. 627-649.

Hage M., Leroy P., *Stakeholder Participation Guidance for the Netherlands Environmental Assessment Agency*, Nijmegen, NEEA Bilthoven/Radboud University Nijmegen, 2007.

Hage M., Leroy P., Petersen A.C., "Stakeholder Participation in Environmental Knowledge Production", *Futures*, 42, 2010, pp. 254-264.

Hajer M., *The Politics of Environmental Discourse: Ecological Modernization and the Policy Process*, Oxford, Oxford University Press, 1995.

Hajer M.A., Wagenaar H. (eds.), *Deliberate Policy Analysis – Understanding Governance in the Network Society*, Cambridge, Cambridge University Press, 2003.

Healey P., "Collaborative Planning in a Stakeholder Society", *Town Planning Review*, 69, 1, 1998, pp. 1-21.

Healy S., "Risk as a Social Process: the End of 'the Age of Appealing to the Facts'?", *Journal of Hazardous Materials*, 86, 2001, pp. 39-53.

Held D., "Catastrophe in the Making: Globalization and the Environment", in Held D., McGrew A. (eds.), *Global Transformations: Politics, Economics and Culture*, Stanford University Press, 1999, pp. 376-413.

Hisschemöller M., Hoppe R., "Coping with Intractable Controversies: the Case for Problem Structuring in Policy Design and Analysis", *Knowledge and Policy*, 8, 4, 1996, pp. 40-60.

Holländer K., Leroy P., "From Scepticism to Good Practices and Tough Challenges", in Thompson Klein J., Grossenbacher-Mansuy W., Häberli R., Bill A., Scholz R.W., Welti M. (eds.), *Transdisciplinarity: Joint Problem Solving among Science, Technology and Society*, Berlin/Basel, Birkhauser Publisher, 2001, pp. 217-235.

Irvin R.A., Stansbury J., "Citizen Participation in Decision Making: Is It Worth the Effort?", *Public Administration Review*, 64, 1, 2004, pp. 55-65.

Irwin A., *Citizen Science: a Study of People, Expertise and Sustainable Development*, London, Routledge, 1995.

Jasanoff S., *The Fifth Branch: Science Advisers as Policy Makers*, Cambridge MA, Harvard University Press, 1990.

Jessop B., "The Rise of Governance and the Risks of Failure: the Case of Economic Development", *International Social Science Journal*, 50, 155, 1998, pp. 29-45.

Jessop B., "Governance Failure", in Stoker G. (ed.), *The New Politics of British Local Governance*, Basingstoke, Macmillan, 2000, pp. 11-32.

Joss J., Bellucci S. (eds.), *Participatory Technology Assessment: European Perspectives*, London, CSD, 2002.

Kasemir B., Jäger J., Jaeger C.C., Gardner M.T. (eds.), *Public Participation in Sustainability Science. A Handbook*, Cambridge, Cambridge University Press, 2003.

Kasemir B., Schibli D. *et al.*, "Involving the Public in Climate and Energy Decisions – Need for Public Support", *Environment*, 42, 2000, pp. 32-42.

Kates R. W. *et al.*, "Sustainability Science", *Science*, 292, 2001, pp. 641-642.

Keller R., Poferl A., "Habermas Fightin' Waste: Problems of Alternative Dispute Resolution in the Risk Society", *Journal of Environmental Policy and Planning*, 2, 2000, pp. 55-67.

Keohane R.O., *Power and Governance in a Partially Globalized World*, London, Routledge, 2002.

Klüver L., Nentwich M., Peissl W. *et al.*, *EUROPTA: European Participatory Technology Assessment. Participatory Methods in Technology Assessment and Technology Decision-Making*, Copenhagen, The Danish Board of Technology, 2000.

Kooiman J., "Societal Governance: Levels, Modes, and Orders of Social-political Interaction", in Pierre J. (ed.), *Debating Governance – Authority, Steering, and Democracy*, Oxford, Oxford University Press, 2000, pp. 138-164.

Kooiman J., "Governance: a Social-Political Perspective", in Grote J.R., Gbikpi B. (eds.), *Participatory Governance – Political and Societal Implications*, Opladen, Leske & Budrich, 2002, pp. 71-96.

Leroy P., "Environmental Politics, Participation and Political Inequality", in Draetta L., Lai F. (eds.), *Naturalia. Sciences sociales et environnement: entre cadres théoriques et approches empiriques*, numéro thématique d'*EUROPAEA*, VIII, 1-2, 2002, pp. 153-167.

Magnette P., "European Governance and Civic Participation: beyond Elitist Citizenship?", *Political Studies*, 51, 2003, pp. 144-160.

Nowotny H., Scott P., Gibbons M., *Re-thinking Science: Knowledge and the Public in an Age of Uncertainty*, London, Polity Press, 2001.

Pearson P., "Political, Epistemological, Ecological and Spiritual Dimensions of Participation", *Studies in Cultures, Organizations and Societies*, 4, 1998, pp. 147-167.

Pellizoni L., "Democracy and the Governance of Uncertainty: the Case of Agricultural Gene Technologies", *Journal of Hazardous Materials*, 86, 2001, pp. 205-222.

Pellizoni L., "Uncertainty and Participatory Democracy", *Environmental Values*, 12, 2003a, pp. 195-224.

Pellizoni L., "Knowledge, Uncertainty and the Transformation of the Public Sphere", *European Journal of Social Theory*, 6, 2003b, pp. 327-355.

Pierre J. (ed.), *Debating Governance: Authority, Steering and Democracy*, Oxford, Oxford University Press, 2000.

Pløger J., "Public Participation and the Art of Governance", *Environment and Planning B*, 28, 2001, pp. 219-241.

Ravetz J., "What is Post-Normal Science?", *Futures*, 31, 7, 1999, pp. 647-653.

Rayner S., "Democracy in the Age of Assessment: Reflections on the Roles of Expertise and Democracy in Public-sector Decision-making", *Science and Public Policy*, 30, 2003, pp. 136-169.

Rhodes R.A.W., *Understanding Governance: Policy Networks, Governance, Reflexivity and Accountability*, Milton Keynes, Open University Press, 1997.

Richards C., Sherlock K., Carter C., *Practical Approaches to Participation*, Aberdeen, Macaulay Institute, 2004.

Roqueplo Ph., *Entre savoir et décision: l'expertise scientifique*, Paris, INRA, 1997.

Scharp L., Richardson T., "Reflections on Foucauldian Discourse Analysis in Planning and Environmental Policy Research", *Journal of Environmental Policy and Planning*, 3, 2001, pp. 193-209.

Schmitter P.C., "Participation in Governance Arrangements: Is There Any Reason to Expect It Will Achieve 'Sustainable and Innovative Policies in a Multi-Level Context'?", in Grote J.R., Gbikpi B. (eds.), *Participatory Governance. Political and Societal Implications*, Opladen, Leske & Budrich, 2002, pp. 51-70.

Siebenhüner B., "Social Learning and Sustainability Science: Which Role Can Stakeholder Participation Play?", *Proceedings of the 2002 Berlin Conference on the Human Dimension of Global Change*, 2004, pp. 76-86.

Smith G., Wales C., "Citizens' Juries and Deliberative Democracy", *Political Studies*, 48, 2000, pp. 51-65.

Spash C.L. (ed.), "Participation, Representation and Deliberation in Environmental Policy", *Environment and Planning*, 19, 4, 2001.

Taylor I.E., "The Genetically Modified Maize Debacle: a Case Study of Policymakers' Failure to Deal with Scientific Uncertainty even after BSE", in Hisschemöller M., Hoppe R., Dunn W.M., Ravetz J.R. (eds.), *Knowledge, Power and Participation in Environmental Policy Analysis*, New Brunswick/London, Transaction Publishers, 2001.

Theys J., "Environmental Governance: from Innovation to Powerlessness", in Grote J.R., Gbikpi B. (eds.), *Participatory Governance. Political and Societal Implications*, Opladen, Leske & Budrich, 2002, pp. 213-244.

Thompson Klein J, Grossenbacher-Mansuy W., Häberli R., Bill A., Scholz R.W., Welti M. (eds.), *Transdisciplinarity: Joint Problem Solving among Science, Technology and Society*, Berlin/Basel, Birkhauser Publisher, 2001.

Torgerson D., "Democracy through Policy Discourse", in Hajer M.A., Wagenaar H. (eds.) (2003), *Deliberate Policy Analysis – Understanding Governance in the Network Society*, Cambridge, Cambridge University Press, 2003, pp. 113-138.

Turnhout E., Leroy P., *Participating in Uncertainty: a Literature Review of Applying Participation in the Delivery of Scientific Policy Advice*, Bilthoven, RIVM (in Dutch), 2004.

Van de Kerkhof M., *Debating Climate Change – A Study of Stakeholder Participation in an Integrated Assessment of Long-Term Climate Policy in the Netherlands*, Amsterdam, Lemma, 2004.

Van de Kerkhof M., Leroy P., "Recent Environmental Research in the Netherlands: Towards Post-normal Science?", *Futures*, 32, 2000, pp. 899-911.

Van Huijstee M., *Businesses and NGOs in Interaction: a Quest for Corporate Social Responsibility*, Utrecht, Netherlands Geographical Studies, 2010.

Van Kersbergen K., Van Waarden F., "Governance as a Bridge between Disciplines: Cross-disciplinary Inspiration Regarding Shifts in Governance and Problems of Governability, Accountability and Legitimacy", *European Journal of Political Research*, 43, 2004, pp. 143-171.

Van Tatenhove J., Leroy P., "Environment and Participation in a Context of Political Modernisation", *Environmental Values*, 12, 2003, pp. 155-174.

Van Tatenhove J., Arts B., Leroy P. (eds.), *Political Modernisation and the Environment – The Renewal of Environmental Policy Arrangements*, Dordrecht, Kluwer, 2000.

Visseren-Hamakers I., *Partnerships in Biodiversity Governance – An Assessment of their Contributions to Halting Biodiversity Loss*, Utrecht, Netherlands Geographical Studies, 2009.

Wildavsky A., *The Art and Craft of Policy Analysis*, London, Macmillan, 1979.

Wynne B., "May the Sheep Safely Graze?", in Lash S., Szerzynski B., Wynne B. (eds.), *Risk, Environment and Modernity*, London, Sage, 1996, pp. 44-83.

Yearley S., Cinderby S., Forrester J., Bailey P., Rosen P., "Participatory Modelling and the Local Governance of the Politics of UK Air Pollution", *Environmental Values*, 12, 2003, pp. 247-262.

EcoPolis

The **EcoPolis** series is dedicated to the analysis of changes that occur simultaneously in society and the environment when the latter becomes a fundamental concern.

For a long time the environment was defined as being external to society, like the world of nature and ecosystems which serves as a material foundation for social life. Thus the aim of environmental policies has been to "preserve", to "protect" or even to "manage" that which was seen as a sort of infrastructure of our societies. After several decades of environmental policy, nature and the environment have become objects of government and it is seemingly within the same movement that each society models its environment and shapes itself. This dialectic will be central to the series.

Series editor: Marc MORMONT,
Professor at the FUL
(Fondation universitaire luxembourgeoise, Université de Liège, Belgium)

Published Books

N° 16– Cécilia CLAEYS and Marie JACQUÉ (eds.), *Environmental Democracy Facing Uncertainty*, 2012, 185 pages.

N° 15– Josiane STOESSEL-RITZ, Maurice BLANC, Nicole MATHIEU (dir.), *Développement durable, communautés et sociétés. Dynamiques socio-anthropologiques*, 2012, 230 pages.

N° 14– Philippe HAMMAN, Christine BLANC et Cécile FRANK, *La négociation dans les projets urbains de tramway. Éléments pour une sociologie de la « ville durable »*, 2011, 246 pages.

N° 13– Denise VAN DAM, Michel STREITH et Jean NIZET (dir.), *L'agriculture bio en devenir. Le cas alsacien*, 2011, 140 pages.

N° 12– Philippe HAMMAN et Jean-Yves CAUSER (dir.), *Ville, environnement et transactions démocratiques. Hommage au Professeur Maurice Blanc*, 2011, 291 pages.

N° 11– Géraldine FROGER (dir.), *Tourisme durable dans les Suds ?*, 2010, 316 pages.

N° 10– Muriel MAILLEFERT, Olivier PETIT et Sandrine ROUSSEAU (dir.), *Ressources, patrimoine, territoires et développement durable*, 2010, 283 pages.

N° 9– Philippe HAMMAN et Christine BLANC, *Sociologie du développement durable urbain. Projets et stratégies métropolitaines françaises*, 2009, 260 pages.

N° 8– François MÉLARD (dir.), *Écologisation. Objets et concepts intermédiaires*, 2008, 214 pages.

N° 7– David AUBIN, *L'eau en partage. L'activation des règles dans les rivalités d'usages en Belgique et en Suisse*, 2007, 247 pages.

N° 6– Géraldine FROGER (dir.), *La mondialisation contre le développement durable ?*, 2006, 315 pages.

N° 5– Laurent MERMET (dir.), *Étudier des écologies futures. Un chantier ouvert pour les recherches prospectives environnementales*, 2005, 411 pages.

N° 4– Jean-Baptiste NARCY, *Pour une gestion spatiale de l'eau. Comment sortir du tuyau ?*, 2004, 342 pages.

N° 3– Pierre STASSART, *Produits fermiers : entre qualification et identité*, 2003, 424 pages.

N° 2– Cécilia CLAEYS-MEKDADE, *Le lien politique à l'épreuve de l'environnement. Expériences camarguaises*, 2003, 245 pages.

N° 1– Edwin ZACCAÏ, *Le développement durable. Dynamique et constitution d'un projet*, 2002 (2e tirage 2003), 358 pages.

FSC
www.fsc.org
MIX
Papier | Fördert gute Waldnutzung
FSC® C083411

Kolibri 360 GmbH
Ferdinand-Jühlke-Straße 7
99095 Erfurt, Deutschland
produktsicherheit@kolibri360.de